The
Self-Generating Universe

How the Universe Can Create Itself from Nothing

By Robert Amneus

Cartoons and Cover by Rudy Adolfo
Drawings by Author
Photographs by NASA

ISBN-13: 978-0-9823263-0-5
ISBN-10: 0-9823263-0-0

Published by: AST Design
www.BooksofScience.com

Dedicated to the Memory of:

Mary Willa Allen
&
Thomas Axel Amneus

Robert Amneus is an honors graduate of UCLA, having achieved a degree in Electrical Engineering with additional studies in Astronomy and Philosophy. As an Integrated Circuit (IC) Design Engineer, Robert holds numerous patents. Robert has worked on hundreds of design projects, including a memory system for the Apollo Moon Program. As a consultant to major US microprocessor design companies, he helped develop state-of-the-art ICs. As the CEO of an integrated circuit design company for over 15 years, Robert personally developed the company's unique software design tools that allowed it to compete at the leading edge of the semiconductor industry. Over thirty years of experience designing tiny microchips has given Robert a unique insight into the possible electrical nature of the tiny starting universe. As a young child, Robert always wondered about the universe and the reason for its existence. This book is the result of his life-long quest for an answer.

Table of Contents

Introduction:

When alone, have you ever looked around and asked yourself:

> *What are these things around me?*
> *Why are they there?*
> *Why do we exist?*
> *Why do we have a universe of something rather than nothing?*

People have been asking such questions since the dawn of human existence. The reason it is different for us is because *today's* science has given us a keyhole to view the answer.

This book peers deeply into that keyhole. In it, we explore the natural laws that govern the creation and structure of the universe. You will see, step by step, how these natural laws allow the self-creation of the universe. You will discover how the Law of Balance allows the universe to be created from nothing without violating the Law of Conservation of Energy.

An electrical model of space is developed for an early universe composed of only two points. Powerful industrial computer simulation tools are then used to show how this electrical model leads to our observable universe.

The book also includes sections on the creation of galaxies, creation of the solar system, and creation of the elements. We further explore the evolution of matter up to the starting point of life.

You will peer through the keyhole of science and discover why the universe exists.

Notes to Reader:

Each section of this book has been made as complete and self-contained as possible. Thus, you may skip to any section without understanding all of the preceding sections. This has resulted in the necessary repeating of some material. This repetition serves to reinforce concepts that may be new to some readers.

We have used the concepts of a particle and of a point particle. For those who do not believe in the physical existence of particles, the wave concept of matter can be substituted equally well to the same end.

Conventional wisdom tells us not to use equations in a book because you will lose your readers. This may indeed be correct advice, but since the universe is governed by mathematical laws, we were compelled to use some equations. If you do not understand the equations, don't feel bad—you are in good company. Most physicists and engineers use these equations to solve problems, yet they are at a loss to understand why nature chose these particular laws to govern the cosmos. We are all limited to a partial understanding, so just give it a try.

The mathematics in the body of this book has been kept to a minimum. Any math beyond simple algebra is in the Appendix, i.e. derivations of some of the fundamental equations of nature. These derivations are presented so that a person with no background in calculus should still be able to follow the derivation and perhaps even gain an appreciation of the elegance and power of calculus.

The essence of calculus is to examine things at the very minute level. In doing so, small error terms in the mathematics become zero, revealing an exact result. When the distance, x, is examined by calculus at the very minute level, this tiny distance is called "dx". This distance is as close to zero as possible without actually being zero. In a similar way, the essence of this book is to examine the universe, u, as "du".

Section 1: Perspectives

Our Time in World History

When Ben Franklin invented the lightning rod in 1752, churches refused to use it since this challenged the divine origin of thunder. As the lightning rod was accepted and put to use, it became apparent that the "Almighty" was hitting churches and killing bell ringers, while leaving the local brothels untouched. Even so, the Church still failed to use the invention. However, in 1769, the Church of San Nazaro in Italy was struck by lightning, setting off 200,000 pounds of stored gunpowder. This explosion resulted in the destruction of one-sixth of the city and the deaths of over 3,000 people.

The Church did not respond to the deaths of its bell ringers. Ultimately, it required a devastating event to bring the Church to reconsider an ancient belief and accept a simple truth of science.

What would our society be like today if we did not experience the suppression of free thought for over a thousand years during the Dark Ages? Would the Internet have been invented a thousand years ago? Would cancer have been cured over 800 years ago? What price do we all pay by ignoring what science has to say?

We are at a time in world history when science has advanced to the state where we can answer, with a high degree of certainty, the most fundamental question regarding our existence: *How did it all come to be?* Although incomplete, the state of today's scientific knowledge is sufficient to show that some of the ancient beliefs most widely held by our civilization need to be reconsidered. It is time to consider the current knowledge we have gained from science. Science has a lot to say about the origin of the universe. It is time to listen.

The Universe is Energy

In one way, the universe is very simple. It is like a soup made from only one ingredient. This one ingredient is energy. Look around you. Any material object that you see is very concentrated energy. This book is energy. Your body is energy. The earth is energy. The sun is energy. The stars and galaxies are energy. Einstein's famous equation, $E=mc^2$, tells us that the mass of any object is highly concentrated energy.

Let's now consider the nature of energy in greater detail.

Most of the energy available to us on earth is derived from the sun. The sun is a nuclear reactor that converts hydrogen atoms into helium atoms. In doing so, some leftover energy is radiated into space. This solar energy sustains all life on earth. The solar energy heats the ocean waters and forms rain. This rain forms rivers that we can convert into hydro-electric energy to power our homes. Plant leaves are small solar collectors that convert sunlight into chemical energy. By eating the plants, we transfer this chemical energy to our bodies. This same chemical energy, bottled up in ancient forests, is released in the burning of oil and coal.

The atoms in our bodies are miniature batteries that store nuclear energy. Like a battery, every atom has a plus and a minus terminal. The center core of the atom is the plus terminal that contains the protons; the outer region of the atom is the minus terminal that contains the electrons. To gain a better appreciation of how much concentrated energy is bundled up in these atomic batteries, consider this example: just one gallon of gasoline has enough *nuclear* energy to power an automobile to drive continuously at 60 miles per hour for over 100,000 years! This is longer than the recorded history of human beings. Alternately, the *chemical* energy in this same gallon of gasoline would only power the automobile for less than an hour. Perhaps in the future we will learn how to tap into this unlimited source of nuclear energy.

We learn about the rest of the world through our senses. Our senses are all energy receptors. Our ability to enjoy beautiful music results from sensing the sound energy in the vibrating air. Our vision of a majestic mountain is the result of light energy received from the mountain. Touch provides us with the ability to sense the binding energy between tiny atoms, and the temperature we feel is a result of our ability to sense the rate of thermal vibration of these atoms.

Energy exists in two fundamental forms. Understanding these two forms of energy is the key to unlocking the mystery of creation. One form is energy of motion, called kinetic energy. A car traveling down the highway has kinetic energy. The other form is a negative energy called potential energy. Potential energy is energy due to the position of an object. A bird in flight has gravitational potential energy due to its altitude. This gravitational potential energy is converted into kinetic energy as the bird dives towards the earth. The water poised at the top of a waterfall has gravitational potential energy. As the water falls over the top, the gravitational energy decreases while the kinetic energy increases.

The energy of mass and gravitational potential energy are very closely related. Gravitational energy does not exist without mass. Both are needed. This marriage of mass and gravity is fundamental to the creation of the universe. The mass provides the positive energy of the universe, while the gravitational potential energy provides a balancing negative energy.

It is this balance of mass energy with gravitational energy that allowed the universe to come into existence. Let's start to understand this. Consider the following picture, where an object is lifted above the surface of the earth and is then allowed to fall.

A Falling Object in the Earth's Gravity and the Key Concept of Gravity as Negative Energy:

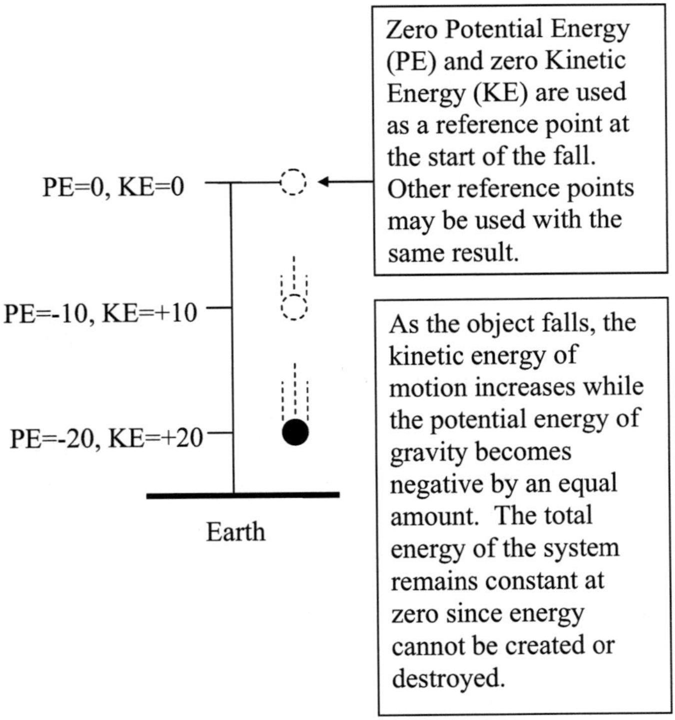

As the object in the above picture falls, the gravitational potential energy becomes more negative, while the kinetic energy becomes more positive by an equal amount. At every point during the fall, the sum of the kinetic energy and potential energy remains at zero. In further sections, we will explore the importance of mass as positive energy—balanced by the negative gravitational potential energy—as a way for the universe to create itself from nothing.

The Law of Conservation of Energy

The Law of Conservation of Energy is a pillar of modern physics. This law states that energy cannot be created or destroyed. Something cannot be created from nothing, and something that exists cannot be turned into nothing. Take a pound of iron and try to destroy it. You can grind it up into small pieces. You can heat it until it turns to vapor—but you cannot convert it into nothing. In a similar way, you cannot take nothing and create a pound of iron. If this law is true, then how did the universe come into existence? If the universe was created from nothing, then wouldn't this be a violation of the Law of Conservation of Energy?

One way out of this dilemma is to declare that the energy of the universe has always existed. In this case, we are conveniently moving the creation into the infinite past. However, the universe provides us with strong evidence that it has not been around forever. For example, if the universe had existed forever, then we would expect the supply of hydrogen that fuels young stars to have been exhausted a long time ago. But this is not what we find.

Fortunately, nature provides us with a simple and elegant way out of this dilemma. In the previous section we presented the concept of the negative potential energy of gravity being balanced by positive kinetic energy. We also pointed out that mass is a form of energy that is tied to the properties of gravity. In the Appendix, we show mathematically how nature seems to require a perfect balance of the positive energy of mass by the negative gravitational energy of space itself. In this case, the total energy of the universe adds up to exactly zero. The positive energy in that pound of iron is balanced by an equal amount of negative gravitational energy somewhere in the universe. We can create that pound of iron from nothing—as long as we create an equal amount of balancing negative energy at the same time. In this way the Law of Conservation of Energy is not violated, since the energy before the creation is zero and the energy after the creation is still zero.

The Simple Circle Shows Us That Nature Makes the Rules

Where did the laws of nature come from? Were they designed by an intelligent being to allow our existence? Or is there another explanation?

Let's start our investigation by first examining one of the most important constants of nature, Pi (π). We know from school that Pi is equal to the circumference of a circle divided by its diameter. Teachers tell us that its value is about 3.14. We know that π is a fundamental constant of nature that determines the properties of a circle, a sphere, and shows up in many of the laws of physics. Let's ask ourselves: Where did this constant of nature come from?

The following picture shows the meaning of π:

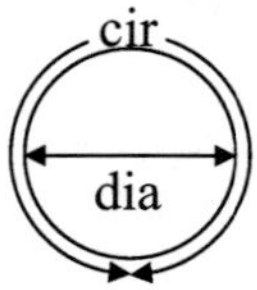

For any size circle:
π = cir/dia =3.14159…

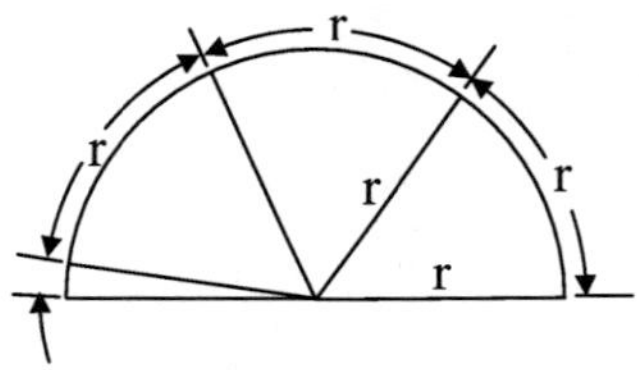

Pi (π) is also the number of times the radius, r, will fit into the arch of a half circle. Pi (π) is an example of a constant of nature that cannot be changed in value.

After studying the above picture, it is clear that π is a constant of nature that is forced to be what it is. Take a tape measure and determine the value for yourself. Mathematicians have determined the value of π to an accuracy that would fill a book if written down. Other intelligent beings in the universe would derive exactly the same value for π. It is just something that *is*. It cannot be changed by even the will of the greatest super being. In the case of π, nature has the ultimate power, not the super being. A super being could not design a universe with a different value of π. Nature has the upper hand.

Could it be that all laws of nature are like π? Let's investigate further by using π to extend our knowledge of the universe.

In the Appendix, the equation for the area of a circle is mathematically derived from the properties of π. We prove that the area of a circle must be:

$$A=\pi r^2$$

No other choice is possible. Again, with the area of a circle, nature set the rules.

In the Appendix we mathematically determine that the volume of a sphere must be:

$$V = \frac{4}{3}\pi r^3$$

Any spherical object must obey this relationship. So far we have determined several properties of nature that are just required to be what they are and cannot be different.

It is interesting that π finds its way into many of the laws of physics. In the Appendix, we show that the surface area of a sphere must be:

$$\textit{Surface area of sphere} = 4\pi r^2$$

Coulomb's Law is a fundamental equation of nature that describes the forces between electrical charges. Coulomb's Law helps to determine the properties of the atoms that make up our body. The surface area of a sphere can be found in this important law of nature:

$$F = \frac{q_1 q_2}{\varepsilon_0 4\pi r^2}$$

It is also interesting that π finds its way into Einstein's famous equation. Below, we rewrite Einstein's equation by replacing c^2 with the more fundamental electrical constants of nature:

$$E=mc^2 = \frac{m}{\varepsilon_0 \mu_0}$$

Where: $\mu_0 = \frac{4\pi}{1E7}$ Henry/meter

We see that the constant π plays a key role in the properties of space as well as in the basic laws of nature that determine the behavior of our universe.

Nature alone has the mysterious power to set many of the properties around us. In this sense, nature makes the rules! The will of the greatest super being cannot change these fundamental properties of nature.

How Nature Transfers Energy

Nature uses the sinusoidal waves described by trigonometry to transfer energy from one location to another. Some examples of such energy transfer include:

- The ripples caused by a rock dropped in a still pond
- The light energy received from a distant star
- Radio waves
- The AC power delivered to our homes
- The combination of sinusoidal pressure waves generated by a violin

- Production of both electrons and protons from colliding sinusoidal electromagnetic waves

In the previous section and in the Appendix, we have mathematically proven that many of the geometric properties of space depend on the constant π, with no other choice being possible. In a similar way, the next figure shows us how the properties of the sinusoidal wave are fundamentally tied to the properties of the circle and π. Is any other choice possible?

The Relation between π, the Circle and the Sinusoidal Wave

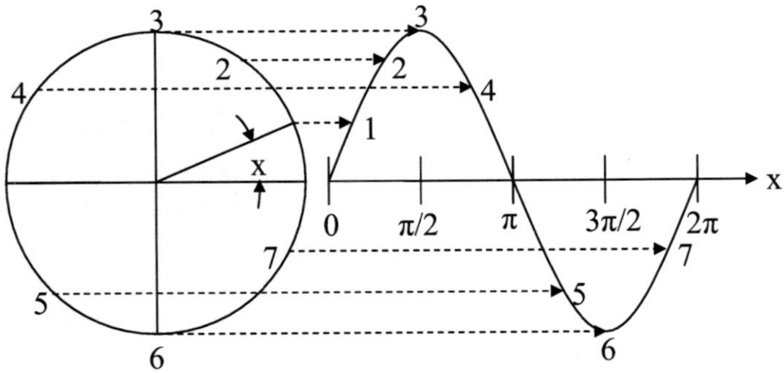

The Inverse Square Laws of Force

Nature provides two fundamental forces, both of which act across very large distances. One of these forces is the electrical force between charged particles. The other is the gravitational force between objects that have mass. Both of these forces are fundamental to organizing a universe that allows us to exist.

The force of gravity organizes the galaxies, the solar system, and the spherical shape of the planets and sun. Gravity causes the stars to glow at high temperatures, resulting in the conversion of simple elements that formed in the early universe into heavier elements that later made life possible.

The electrical force determines the properties of the different atoms that allow life to exist. As an analogy, think of the Golden Gate Bridge. The weight of the bridge deck is supported by cables that run along the top of the bridge. It is the electrical force between the atoms in this cable that allows the bridge to stand. Any cross section of this cable is only as strong as the force of attraction between the atoms within the cross section. When we drive across the bridge, we put a lot of faith in the electrical force.

Both of these force laws are very similar in that they are both represented mathematically by what is called an "inverse square law relationship." So far, we have shown that the properties of the circle and sphere are determined by nature and are unchangeable. Was this inverse square law relationship designed, or is it a property of nature similar to the constant π that cannot be changed?

To answer this question, let's first review the nature of a typical inverse square law equation used to define a force.

Math 101 Review of the Inverse Square Law

$$F = a\frac{bc}{d^2}$$

On the left side of the equal sign is the force, F. The terms to the right of the equal sign tell us how to calculate the value of the force.

The term bc in the above equation tells us to multiply the value of b times the value of c.

The term d^2 in tells us to multiply d times d. The d term is said to be squared, since the area of a square with a length of one side equal to d, is d times d.

The terms $\frac{bc}{d^2}$ tell us to divide the result of b times c by the result of d times d.

The a term is multiplied by the result of the $\frac{bc}{d^2}$ calculation.

The abc terms are in the numerator of the equation.

The d^2 term is in the denominator of the equation.

Math Review Continued

Note that as the numerator increases in value, the force also increases in value. As the denominator term decreases in value, the force increases. The force is said to have a direct relationship with the value of the numerator and an inverse relationship with the value of the denominator. As d approaches a value of zero, the force approaches a value of infinity.

Coulomb's Law is an inverse square law that defines the electrical force between two charges, q_1 and q_2, separated by a distance, d:

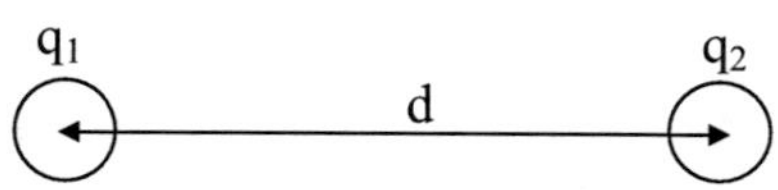

This force is written as the following inverse square law:

$$F = \frac{q_1 q_2}{4\pi\varepsilon_0 d^2}$$

Where:

F = Force between charge q_1 and charge q_2
q_1 = charge of object 1
q_2 = charge of object 2
π = 3.14159
d = distance between the center of each charge
ε_0 = Permittivity constant (This is a fundamental constant related to the electrical properties of space.)

Newton's Law of Gravitation describes the force of attraction between two masses, m_1 and m_2, separated by a distance, d:

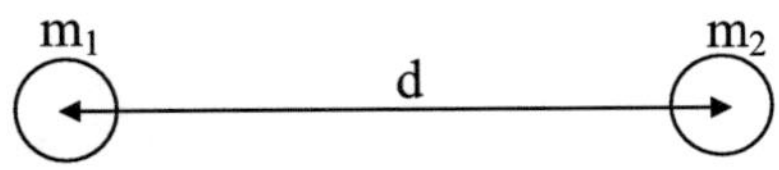

This force is also an inverse square law:

$$F = G\frac{m_1 m_2}{d^2}$$

Where:

F = Force of attraction between mass m_1 and mass m_2
G = Gravitational Constant
m_1 = mass of object 1
m_2 = mass of object 2
d = distance between the center of each mass

Note that these forces are of a very similar form. Both forces are defined by a constant that is multiplied by the mass or charge of two objects and divided by the square of the distance between the objects:

$$F = \text{Constant}\frac{(\text{mass or charge1}) * (\text{mass or charge2})}{\text{distance squared}}$$

Again, this is called a square law since the distance between the objects is squared. This is called an inverse square law since the force changes inversely related to the distance. As the distance between the objects becomes smaller, the force becomes larger. As the distance between the objects is cut in half, the force increases by a factor of four, due to the squared value in the denominator.

According to the inverse square law, an early universe composed of only two point charges or two point masses would have an infinite force between them as the distance between them becomes zero. Could the force of the Big Bang explosion have been generated from the inverse square law relationship of these two fundamental forces? Could the universe be a result of a simple division by zero, resulting in an infinite force to power the creation? Isn't it interesting that both the force laws that have the capability of shaping the large-scale properties of the universe also have the power of a division by zero when the distance between objects in the early universe is very small?

Why are such important forces defined by the inverse square law? Why wouldn't the exponent be 2.00001 instead of exactly 2? Were these inverse square laws, which determine the properties of our universe, a result of design by a super being? Were they just an accident or are they required to be as they are?

To better understand the inverse square law, let's consider an analogy with the law of luminosity, which also obeys the inverse square law. The next figure shows a light source that spreads out uniformly into space, similar to gravity and the electric force. Region B is twice the distance of region A from the light source. It can be clearly seen that one of the B regions receives one-fourth the luminosity that region A receives. You can also easily see that doubling the distance from the light source cuts the luminosity by one-fourth, thus obeying the inverse square law.

How the inverse square law works:

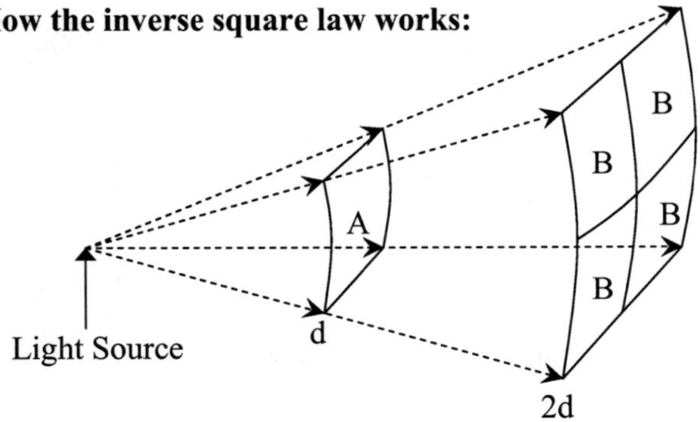

Moving twice the distance from the light source reduces the light intensity by a factor of four.

The picture above reveals that the inverse square law is a result of the property of a sphere. As we show in the Appendix, the surface area of a sphere obeys the equation:

$$\text{Area}=4\pi r^2$$

Where:

r = radius of the sphere

From this simple equation, we can see that, as the radius of the sphere doubles, the surface area increases by a factor of 4. This causes the luminosity per area to drop by a factor of 4, as is expected from an inverse square law.

Like the constant π and like other properties of geometry, the inverse square law has no choice. It is a fundamental property. Nature makes the rules and they cannot be changed by the will of the greatest super being.

The Creation Power of a Divide by Zero

In this section, we show that nature has the tools needed to provide the energy for its own creation.

Nature provides three energy laws that take the form:

Energy = a number greater than zero divided by a distance

Remember, any number divided by zero becomes infinity. Thus, nature provides a very simple way to generate the very large energies needed for self-creation.

These energy laws include the gravitational energy between objects of mass, the electrical energy between objects having charge, and the energy of the electromagnetic wave. Light is an example of such a wave. In the cases of mass and charge, the distance is the space between the objects. The smaller the space, the greater is the energy. In the case of electromagnetic energy, the distance is the space between the adjacent crests of the vibrating waves. In all three cases, as the distance approaches zero, the energy approaches infinity. Thus, nature can provide a lot of energy to power a very small universe at the outset. The smaller the beginning universe, the more energy is available for its self-creation.

The following graph demonstrates this energy relationship:

The power of a divide by zero

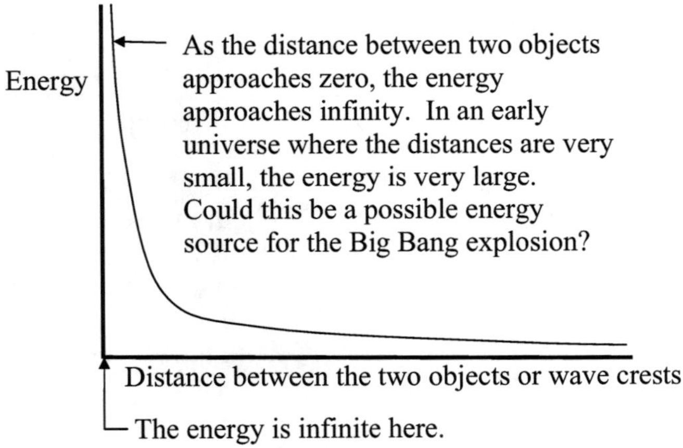

Thus, nature provides a mechanism where the smaller the distance between objects, the greater the energy for self-creation. Also, the smaller the size of the early universe, the greater is the electromagnetic energy in that universe. This seems counter-intuitive because something small has more energy than something large.

In the section on Olbers' Paradox, we will present evidence that the universe cannot be infinite. If true, then this is evidence that nature does not do a true divide by zero, since this would result in an infinite universe. Whether or not nature can do a true divide by zero, it seems certain that nature has the energy tools needed for self-creation.

Given Enough Time, All Unlikely Events Will Occur

A universe that can exist for a very long time provides nature with very strong creative powers.

In the Appendix, we have derived the laws of probability that tell us how much time is needed for a low-probability event to occur. These laws of probability tell us that an event with a very small chance of occurring will occur many times, given enough time. For example, how long would it take for a self-replicating molecule to come into existence? The probability equations tell us that as long as the chance of a self-replicating molecule coming into existence is greater than zero, then all we need is enough time for this event to occur in nature.

For example, how long will it take for a monkey typing randomly at a keyboard to type "hi"? If the chance of hitting the correct key for each character is one in a hundred, then the probability of typing these two characters in a given try is:

$$(1/100)^2 = 0.0001$$

The law of probability tells us that if the monkey makes 100,000 typing attempts, then we can be 99.9955 percent confident that "hi" will be typed at least one time.

How long will it take for a monkey typing randomly at a keyboard to create Shakespeare's *Hamlet*? Sounds impossible, doesn't it? But if this monkey could type for an infinite amount of time, then the laws of probability require that Shakespeare's *Hamlet* will not only be typed once, but will be created an infinite number of times (along with many other unexpected creations).

Since nature has this power of creation, it seems less surprising that the universe, the DNA molecule, and humans exist. The stable creations of nature remain and the unstable creations are lost to history.

Evidence for a Finite Universe

Olbers' Paradox

Why is the night sky dark? A simple calculation presented in the Appendix shows that an infinite universe filled uniformly with galaxies should be glowing with the brightness of a star. But this is not what we see when we view the night sky.

The following picture shows a very large section of the universe where we are at the center. Also shown are two rings similar to the layers of an onion. The outer ring has a larger volume than does the inner ring and will, on average, contain more galaxies. However, the outer ring is at a greater distance from us, so the luminosity reaching us from it will be decreased. Mathematical analysis presented in the Appendix shows that the luminosity reaching the earth from each ring is the same. When the sum of the luminosity reaching the earth from all possible rings is added up for an infinite radius, the luminosity reaching the earth becomes infinite. It can be argued that the universe may have dust or other dark matter that will block the light from the outer rings. However, such matter would heat up to thousands of degrees and glow like the rest of the matter, all roasting in a big cosmic oven.

As previously discussed, Olbers' Paradox is evidence that nature does not do a true divide by zero in the energy equations since this would result in an infinite universe.

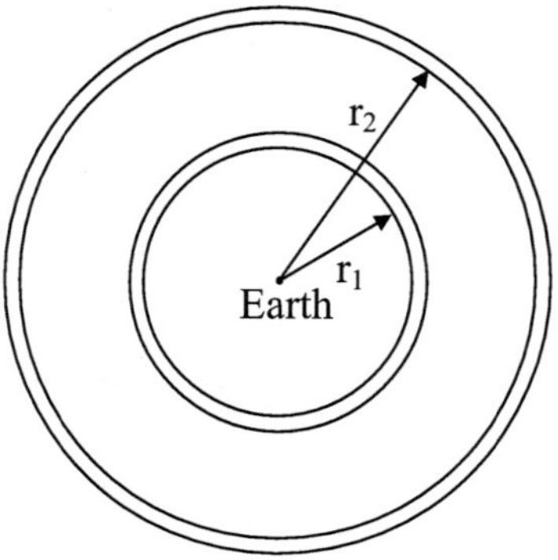

The luminosity reaching the earth from the ring at r_1 and r_2 is equal. If the luminosity reaching the earth from all possible rings is added up for an infinite radius, then the total luminosity becomes infinite. In this case, the night sky should be as bright as a star.

When we look up at the night sky and see mostly darkness, this is evidence that we live in a finite universe.

Hubble's Law

In the 1920s, Edwin Hubble was able to measure the velocity of many galaxies as well as the distance of these galaxies from earth. He found that the galaxies were moving away from the earth. He determined that the further the distance from earth, the faster the galaxies were moving away. This provided some of the first evidence that the universe was expanding. The galaxies are like raisins in a loaf of bread; the raisins become more separated as the bread expands during baking. The further the raisins are separated, the faster they move away from each other.

The following is a graph of Hubble's Law:

Hubble's Law

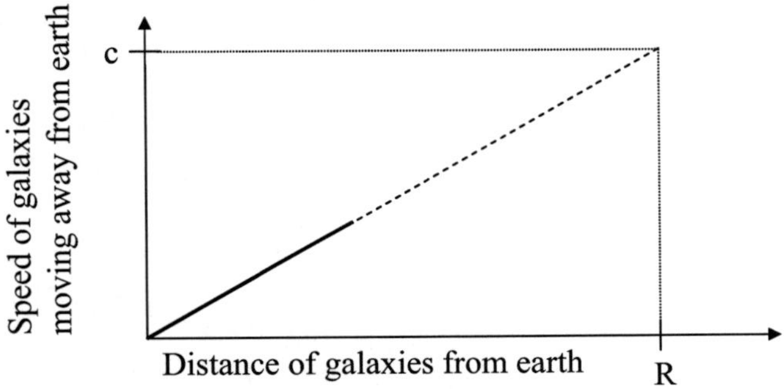

c = speed of light ($3x10^8$ meters/second)
R = Hubble Radius of Universe (14 billion light years)
H = Hubble's Constant = c/R

Hubble's Law also provides us with an approximation for the age of our universe—about 14 billion years. This age is in approximate agreement with the age estimates of old stars based on observation of the rate of radioactive decay of thorium and uranium 238.

If the universe is expanding, then, at an earlier time, the galaxies were closer together. As we move back in time, we come closer to the start of the Big Bang.

Section 2: The Guiding Principles for the Self-Creation

Why Would a Universe of Nothing be Unstable?

A universe of nothing is not that difficult to imagine:

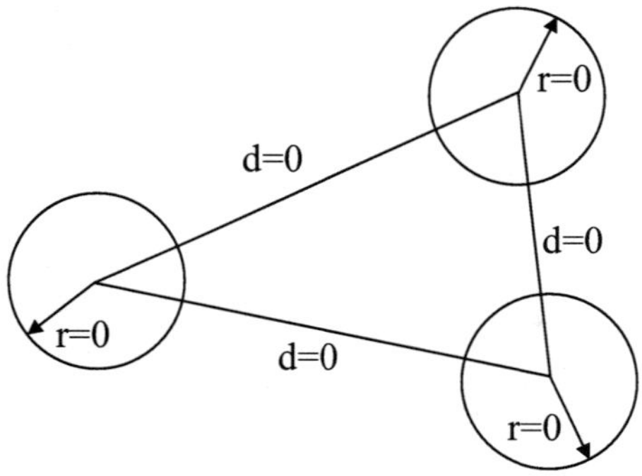

A universe of nothing can be imagined as spheres of zero radii with zero spacing between the spheres.

The start of the universe can be imagined as any one of these spheres having its radius change from zero to a value slightly greater than zero.

Let's consider what happens when a trigger event causes one of these spheres to undergo a small increase in radius. Space will now only exist within the radius of this small sphere—nothing will exist outside. The pressure ratio between the region inside of the sphere and the region outside of the sphere is close to infinite. This great pressure difference causes the radius of the sphere to rapidly increase, resulting in a "Big Bang" explosion.

Imagine a partially filled balloon placed in a vacuum chamber. As the air is removed from the chamber, the balloon must expand due to the pressure difference. If the space outside of the balloon could also be removed and replaced by "nothing," then no pressure could exist outside of the balloon to limit the expansion. The balloon would now be expected to expand much more rapidly.

We will later show how the radius of a sphere can increase from zero and thus give birth to all of the energy in our universe without violating the Law of Conservation of Energy.

It is not difficult to understand how a universe of "nothing" could be unstable. In this case, the "Big Bang" event cannot be stopped. The slightest fluctuation will be enough to trigger the creation of the universe. We will later show how the physics of virtual particle generation require the trigger event that sets off the Big Bang explosion.

The only way to stop the universe from self-creation may be a deliberate intervention from a super being that is beyond physics. This super being would be kept continuously busy, stopping all of the virtual particle trigger events that threaten to set off the Big Bang explosion. Could the fact we have a universe actually be evidence *against* design by an all-knowing super being?

The Smaller the Universe, the More Unstable It Is

Let's consider an imaginary universe that starts out as only two tiny point particles; these particles are far smaller than the tiny electron. If these particles start out close enough to each other then they will have enough gravitational energy to produce all of the galaxies and other energy observed in our universe. In fact, as the distance between these particles approaches zero, the gravitational energy of the particles approaches infinity. This may seem impossible, but the laws of physics tell us that this is true. The smaller the distance between the early particles of the universe, the more energy is available to power the Big Bang explosion. Let's see how this works.

Newton's Law of Gravitation describes the force of attraction between two masses, m_1 and m_2, separated by a distance r:

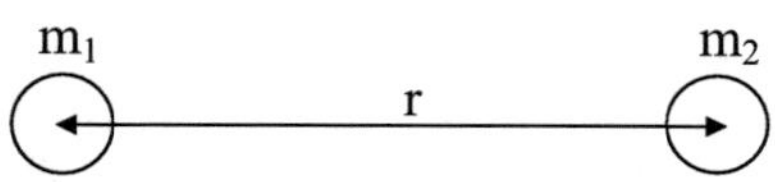

This force is written as:

$$F = G \frac{m_1 m_2}{r^2}$$

Where:

- F = Force of attraction between mass m_1 and mass m_2
- G = Gravitational Constant
- m_1 = mass of object 1
- m_2 = mass of object 2
- r = distance between the center of each mass

The energy of the two masses shown above is the force times the distance of separation:

Energy = (force)*(distance)

At a distance of separation, r, the energy between the two particles becomes:

$$Energy = G\,\frac{m_1\,m_2}{r}$$

We can now graph the energy of our imaginary two-point-particle universe versus the separation distance of the particles:

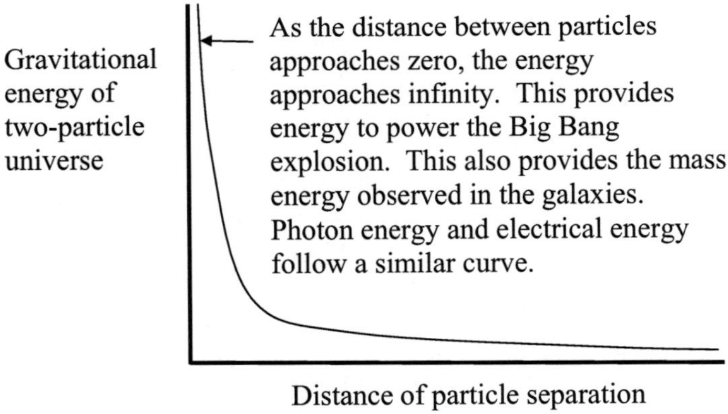

Thus, the Law of Gravitation is one of the powerful clues to understanding how nature provides the energy needed to create the universe. We just need two mass point particles to start out very close to each other. Since gravitational forces are attractive, the two particles in our imaginary universe will be propelled toward each other while gaining more and more energy. They will rebound from the collision with near-infinite energy, resulting in a terrific explosion. This is very similar to what happens in a super nova star when the core collapses inward, setting off a massive outward explosion.

However, not all of the energy of the early universe is in the form of gravitational energy. The second energy source in the early universe that we need to consider is the electromagnetic energy of photons. The energy of a single photon is given by the equation:

$$Energy = \frac{h}{\lambda}$$

Where:

h = Planck's constant (6.63E-23 joules-seconds)

λ = the wavelength of the photon

An early universe of a very small size would be restricted to photons of very small wavelengths. These photons would gain near-infinite energy as the photon wavelength approaches zero. Thus the energy of a single photon with a very short wavelength would contain enough energy to power the Big Bang explosion and create the energy of the observed universe.

The third energy law that we need to consider is the electrical energy between charged particles in the early universe.

Coulomb's Law describes the force between two charges, q_1 and q_2, separated by a distance, r:

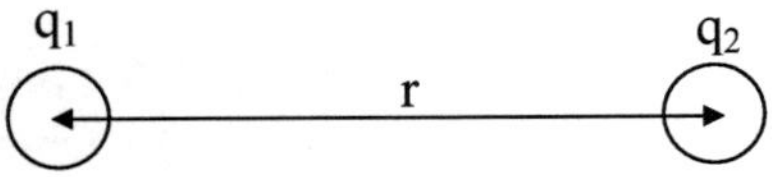

This force is written as:

$$F = \frac{q_1 q_2}{4\pi \varepsilon_0 r^2}$$

Where:

F = Force between charge q_1 and charge q_2
ε_0 = Permittivity constant of space
q_1 = charge of object 1
q_2 = charge of object 2
$\pi = 3.14159$
r = distance between the center of each charge

The energy of the two charges shown above is the force times the distance of separation:

Energy = (force)*(distance)

At a separation distance of r, the electrical energy between the two particles becomes:

$$Energy = \frac{q_1 q_2}{4\pi \varepsilon_0 r}$$

This result is very similar to the result for gravitational energy. As the distance between the charged particles approaches zero, the electrical energy approaches infinity.

It seems that nature is trying to tell us something. In a very small early universe, the gravitational, the photon, and the electrical energies all approach infinity as the radius of the early universe comes very close to zero. The smaller the radius of the early universe, the more compressed energy it has—and the more unstable it is. A universe with a radius of 2r is more unstable than a universe of 3r. A universe with a radius of r is more unstable than a universe of 2r. A universe of zero is the most unstable universe possible.

Could it be that the reason we have a universe of "something" rather than "nothing" is because a universe of "nothing" (zero radius) is unstable?

How Can Something be Created from Nothing?

Recall that energy cannot be created or destroyed. Also recall that mass is just a very concentrated form of energy. Then, how can we have a universe that is full of energy without violating the Law of Conservation of Energy? How can a universe of something come from nothing?

As previously discussed, nature provides a way around this dilemma. As long as the total energy of the universe always adds up to zero, the Law of Conservation of Energy has not been violated. We will now explore this concept of energy balance in greater detail.

To illustrate this concept of energy balance, let's consider a hypothetical universe where negative mass exists. In this universe, the earth would exert a repulsive gravitational force on the negative mass. Below is shown a balance where the left pan has an equal amount of positive mass and negative mass. The negative mass is placed at the bottom of the left pan, since the earth exerts an upward force and we don't want it to fly away. When we balance the masses in this way, we find that both the left pan and right pan have an equal net weight of zero. Mathematically we can state that the mass in the left pan is equal to the mass in the right pan:

$$+M - M = 0$$

The left pan illustrates a process that could create a universe. As long as equal amounts of positive and negative mass are added to the left pan, the net positive mass of the universe can be increased—as long as it is balanced by an equal amount of negative mass. During this creation process, the total of energy of the universe is always zero. We can have a universe without violating the energy conservation law.

Concept of Negative Energy Balanced by Positive Energy

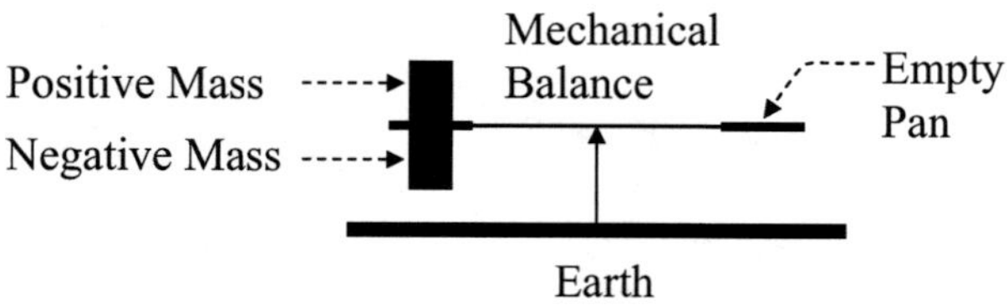

In the above example, we have shown the creation process in terms of positive mass and negative mass in balance. But the universe may also have come into existence from a wave of positive energy balanced by the wave of negative energy. The next picture shows how this could have happened.

The Origin of the Universe from a Balanced Energy Wave

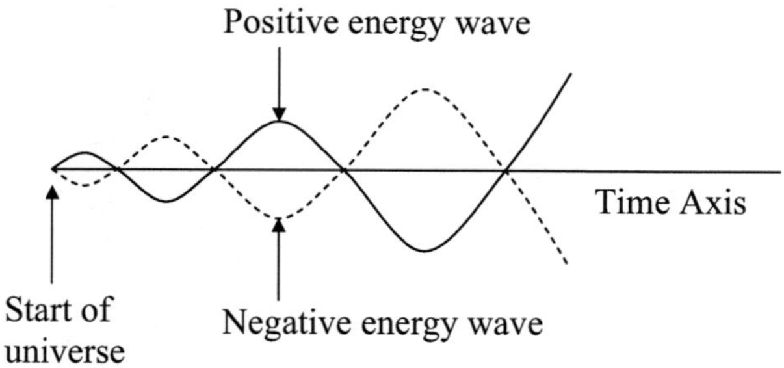

We see that the universe could have started out as a very small wave and grown in both positive and negative energy, as long as the positive part of the wave was always balanced by the negative part at every point in time.

The Law of Balance and the Creation of the Universe

Before going into the creation process in greater detail, let's first review the fundamentals of energy:

- Everything in the known universe is energy.
- Einstein's famous equation, $E=mc^2$, tells us that matter is concentrated energy.
- The Law of Conservation of Energy tells us that energy cannot be created or destroyed. Burning a piece of wood converts the wood into gases and charcoal, but the total amount of energy before and after the fire is unchanged.
- Energy balance allows the creation of the universe without violation of the Law of Conservation of Energy.

The most simple of all truths is the Law of Balance:

$$E=E$$

This self-evident law states that something is equal to itself. In the case of the universe, E is the total positive energy of the universe. At first glance, this equation does not seem to be very profound. But let's see what happens when we do a little bit of algebra. We subtract E from both sides of the equation and replace the right side of the equation with zero:

$$E-E=E-E$$

$$E-E=0$$

The above equation is now more interesting if we consider E to be the total positive energy of the universe and -E to be the total negative energy of the universe. This equation states that the universe can have an amount of positive energy that is exactly balanced by an equal amount of negative energy, resulting in a universe where the total energy is zero.

This equation provides a way for the universe to evolve from nothing without violating the Law of Conservation of Energy. During the creation of the universe, each time a small amount of positive energy is created, an equal amount of negative energy must be created. At all times, the net energy of the universe must be zero.

With this concept of both positive and negative energy, we can now start to understand how the universe might have evolved from nothing without violating the conservation law. Now we can say that in the very distant past, before the universe existed, that the total energy content of the universe is represented by the following equation:

$$0 - 0 = 0$$

At some later time we can imagine an event that caused a small amount of positive and an equally small amount of negative energy to come into existence. We will later show how this event takes place. Using the number 1 to represent a very small amount of energy, this event can be represented by the following equation:

$$1 - 1 = 0$$

At a later point in time we can imagine this same event occurring again in such a way as to add to the previous energy. This event is represented by the following equation:

$$2 - 2 = 0$$

We can now represent an evolution of the universe in the following steps:

$0 - 0 = 0$ (Before the universe existed)
$1 - 1 = 0$ (Universe starts to evolve)
$2 - 2 = 0$
$3 - 3 = 0$ (Universe continues to evolve)
$4 - 4 = 0$
*
*
$E - E = 0$ (Energy state of our present universe)

We will later investigate how the first step, $1 - 1 = 0$, could have taken place in the evolution of the universe and why the second step, $2 - 2 = 0$, must soon follow.

We will later show mathematically how the positive mass energy of the universe, E, can be exactly balanced by the negative gravitational energy, -E, of the universe. The mathematical derivation presented in the Appendix shows that the universe seems to require this balance of energy.

Once you understand how this positive and negative balance works, you will have a renewed sense of wonder when you view many common processes in nature. For example, you will never again view a waterfall in the same way. This energy concept may not be easy to fully grasp, but it will be well worth the effort.

Nature's Way of Triggering the Universe

Quantum Mechanics is the branch of physics that describes the behavior of nature at the very small scale of the atom and subatomic particles. The Heisenberg Uncertainty Principle is a fundamental law of nature that is part of Quantum Mechanics. This principle tells us that perfect knowledge at the atomic level is impossible. For example, to say that an exact point in space has exactly zero energy is not possible, according to the Uncertainty Principle. There is always some probability that non-zero energy in the form of virtual particles will be present at any point in space. This is exactly what nature needs to trigger the universe into existence.

The Heisenberg Uncertainty Principle is shown below in terms of energy and time:

$$\Delta E * \Delta T \geq \frac{h}{4\pi}$$

Where:

ΔE = the uncertainty in energy
ΔT = the uncertainty in time
h = a very small number called Planck's constant

In the above equation, ΔE is multiplied by ΔT. The result of this multiplication must be equal to or greater than a very small number that does not change in value:

$$\frac{h}{4\pi}$$

This equation tells us that an amount of energy, ΔE, can come into existence for a time, ΔT. As the amount of energy that comes into existence becomes smaller, this equation tells us that the time interval for the existence of this energy becomes larger. As ΔE approaches a value of zero, ΔT approaches a value of infinity. This equation tells us that a universe where the total energy sums to zero could come into existence for an infinite time.

According to the Uncertainty Principle, nature does not allow a perfect vacuum to exist—the state of "nothing" is outlawed by nature. Virtual particles are always popping in and out of existence in every region of space. Scientists observe the effects of these virtual particle creation events in laboratory experiments.

One of these experiments is the Casimir effect. Here, two parallel metal plates are placed very close to each other, but do not touch. The small gap between the plates restricts the allowed virtual particle energy levels. The region outside of the plates does not have this restriction. Thus, the small gap in the center has fewer virtual particles hitting the plates than does the region on the outside of the plates. This results in a force that pushes the plates together. This resulting force has been measured in the laboratory, proving that virtual particles are a real phenomenon.

The Casimir Effect

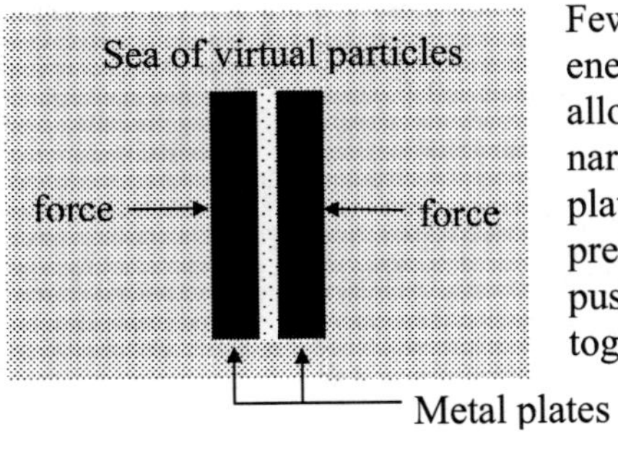

Fewer virtual particle energy levels are allowed to exist in the narrow gap between the plates. This results in a pressure difference that pushes the plates together.

In addition to the Casimir Effect, another discovery known as the Lamb Shift provides additional evidence for the existence of virtual particles. In 1955, Willis Lamb won the Nobel Prize in physics for his discovery of the effect that virtual particles have on the light emitted from the hydrogen atom.

Besides providing the trigger for creation, virtual particles also produce nature's building blocks. These building blocks include the photon (particle of light); the proton and neutron that form the core of the atom; and the electron that forms the outer region of the atom and gives the elements their chemical properties.

Virtual particles always come in pairs—a particle and antiparticle. The antiparticle of the electron is the positron. The positron is the same as the electron except it has a positive charge. The antiparticle of the proton is the antiproton. The antiproton has a negative charge. When a particle combines with its antiparticle, electromagnetic radiation such as light is produced. This process is reversible when electromagnetic radiation comes together to produce particles and antiparticles. Thus, nature allows energy to transition back and forth between electromagnetic energy (such as light or photons) and particles.

Near the start of the Big Bang, energetic photons collided with each other to produce particles along with their antiparticles. This same creation of particles is observed in laboratory experiments when beams of high-energy photons are made to collide. Most of the particles produced in this way will recombine with their antiparticles. Since most of the matter that we observe is composed of protons and electrons rather than their antiparticles, the laws of nature seem to have a bias that favors the production of electrons and protons over their antiparticles.

The universe cools as it expands from the initial Big Bang. After approximately 0.0001 seconds, the universe has cooled enough so that the average photon energy is less than the mass energy of the neutron and proton. At this point, the production of neutrons and protons (and antiprotons) stops. Since electrons only have about 1/2000 the mass of the proton, they are generated for the first four seconds after the start of the Big Bang. At this point, the average photon energy level has dropped below the mass energy of the electron, causing electron and positron production to stop.

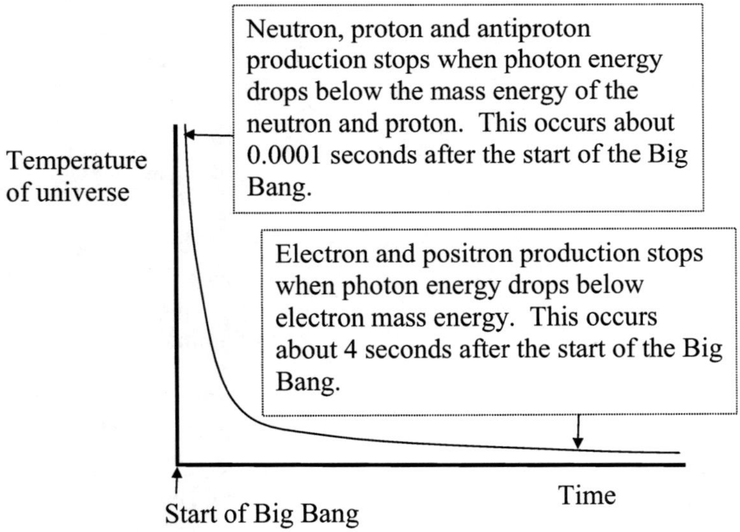

Physicists calculate that the visible matter in the early universe was composed of approximately 74 percent hydrogen, 24 percent helium, and trace amounts of other elements. This concurs with the spectral analysis of the light coming from the most ancient stars. The large planets, Jupiter and Saturn, are composed of mostly hydrogen and, to a lesser extent, helium. The other elements such as carbon, nitrogen, oxygen, calcium and iron necessary to make our bodies were later produced in the hot interiors of stars. The stars converted hydrogen into heavier elements by high-temperature nuclear fusion. While the production of hydrogen occurred in the first few minutes of the universe, the production of these heavier elements required billions of years.

Now let's reconsider the Law of Balance and how the universe could have evolved from nothing with virtual particle generation as the trigger mechanism. Let's rewrite the Law of Balance as:

$$e - s = 0$$

Where:

e = positive energy of the universe in the form of light, heat, mass

s = negative energy of the universe in the form of space itself, including the negative energy due to gravity

Let's assume that the universe starts out as nothing:

$$0 - 0 = 0$$

Now let's consider that a small virtual particle pops into existence in such a way as to remain stable. As the virtual particle comes into existence, an equal amount of space containing the negative energy must also come into existence. The virtual particle is represented by e and the space that contains the particle is represented by -s:

$$e - s = 0$$

The enormous pressure difference between the newly created region of space and the region outside of it causes the newly created region to expand—similar to putting a filled balloon into a vacuum chamber. This rapid expansion of space must be balanced by the creation of additional virtual particles, as shown:

$$2e - 2s = 0$$

Now we have a bigger region of space expanding to fill the empty surrounding space. To balance the expanding space, more virtual particles must come into existence so that the Law of Conservation of Energy is not violated:

$$3e - 3s = 0$$

This process continues until we have the universe of today where the positive energy is exactly balanced by the negative energy in the gravitational field of space:

$$E - S = 0$$

Where:

E = total positive energy of current universe
S = total negative energy of current universe

Recall that the smaller the volume of the universe, the less stable it is as required by the energy laws of photons, gravity and electric charge. As the radius of the universe approaches zero distance, the energy of the particles approaches infinity. This enormous energy between particles crowded into a very small region of space propels the Big Bang explosion.

As the radius, r, of the universe increases, the volume of the universe increases at a much faster rate:

$$\text{Volume} = (4/3)\pi r^3 \quad \text{(Equation for the volume of a sphere)}$$

This rapid increase in volume requires a very rapid increase in the number of virtual particles in order to maintain energy balance. These particles crowded into the small initial volume will generate an enormous heat to propel the Big Bang. The Big Bang is further accelerated by the enormous pressure difference between the newly created region and the nothing surrounding this region. Once the universe of nothing has a virtual particle trigger event, the Big Bang explosion occurs very rapidly. We have a universe of something rather than nothing because a universe of nothing is unstable. Virtual particle creation provides strong evidence that the universe is capable of triggering its own existence.

Section 3: The Properties of Space and the Self-Creation

A massive object causes a curvature of nearby space. This curvature causes the beam of light to follow a bent path.

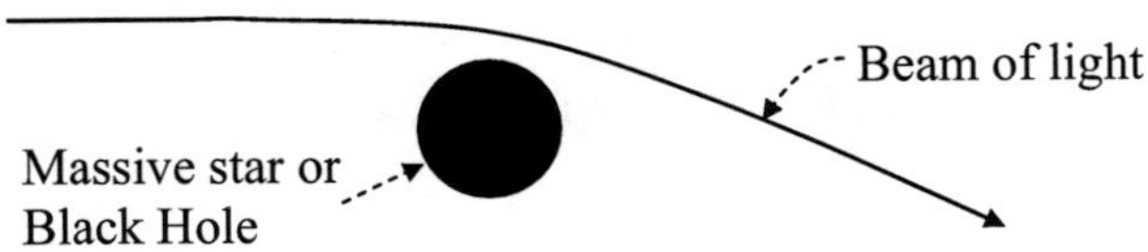

A fundamental relationship exists between mass, gravity and the shape of space. This was predicted by Einstein's General Theory of Relativity and later verified by measurements. Gravity cannot exist without mass and mass cannot exist without gravity. This marriage of mass and gravity is the key to the creation of the universe.

Gravity as Negative Energy and the Law of Balance

Is the human mind even capable of understanding the mysteries of existence? It difficult for the human mind to understand existence, just as it is difficult for a dog's mind to understand the game of chess. The concept of gravity as negative energy is essential to understanding how the universe evolved from nothing. Having touched upon this concept, we will now investigate gravity in greater detail. The following examples will help you form the necessary connections to grasp this key concept.

Consider a bird in flight. As the bird dives toward the ground, its energy of motion increases. Since energy cannot be created or destroyed, where did this energy increase come from? When the bird was flying high up, it had more gravitational energy. As the bird dives toward the ground, its gravitational energy decreases. This decrease in gravitational energy is exactly balanced by the increase in the bird's energy of motion. The bird has gained one unit of energy of motion while losing one unit of gravitational energy. We can mathematically represent the net change of energy as one unit gained in energy of motion minus one unit lost in gravitational energy resulting in a net energy change of zero. Representing this unit of energy as 1, we can write the net energy change as:

$$1-1 = 0 \quad \text{(An example of the Law of Balance)}$$

The first term in this equation is the bird's increase in energy of motion. The second term is the negative energy stored in the gravitational field. The energy of gravity is a negative energy, as any college physics book will confirm.

When we see the bird dive, we tend to see the positive energy increase in the bird's motion. We tend *not* to see the decrease in gravitational energy. If the bird decides to return to its original height, then it will need to burn some of its chemical energy gained from its last meal to gain back the gravitational energy. Energy can change from gravitational energy into energy of motion and then from chemical energy back into gravitational energy. During this process, energy changes from one type to another but is never created or destroyed. Thus energy is always conserved.

These same energy laws that apply to the bird in flight also apply to all material objects in the universe. A collapsing black hole formed by a super-massive star obeys these same energy laws when its mass and gravity are compressed into a very small volume. Could this mass and gravity be compressed out of existence? If so, then the time reversal of this black hole compression would be the creation process.

Let's now consider a rocket ship of mass "m" sitting on the surface of the earth. As the rocket ship is launched from the earth's surface, it gains in gravitational potential energy. As the rocket ship climbs, its gravitational potential energy becomes less negative. If the rocket ship has enough fuel and time, it will eventually approach an infinite distance from the earth. As the rocket approaches an infinite distance, the gravitational potential energy approaches zero. When the rocket was at the earth's surface, it possessed the most negative gravitational energy. As the rocket continued to climb, the gravitational energy continued to increase until the rocket reached an infinite distance. At that point the negative gravitational energy increased to zero.

In the Appendix (See "How the Total Energy of the Universe Sums to Zero"), we show mathematically how much energy is required to move an object of mass "m" to an infinite distance from the universe. It is very interesting that the amount of energy needed is mc^2, in agreement with Einstein's famous equation. This tells us that if the rocket ship in the previous example could use its own mass as rocket fuel, then the mass of the rocket ship would become zero at an infinite distance. The positive mass of the rocket has just disappeared, along with an equal amount of negative gravitational energy.

This is an important clue as to how objects in our universe could have come into existence from nothing. If we could take all material objects in the universe and use their mass as rocket fuel to escape each other's gravitational pull, then the mass of the universe would disappear along with the gravitational energy. We would be left with a universe of nothing. A reversal of this process would create a universe from nothing. A simple energy balance between the positive mass energy and the negative gravitational energy helps to explain how our universe can self-create from nothing.

Moving the Mass of the Universe to Infinity

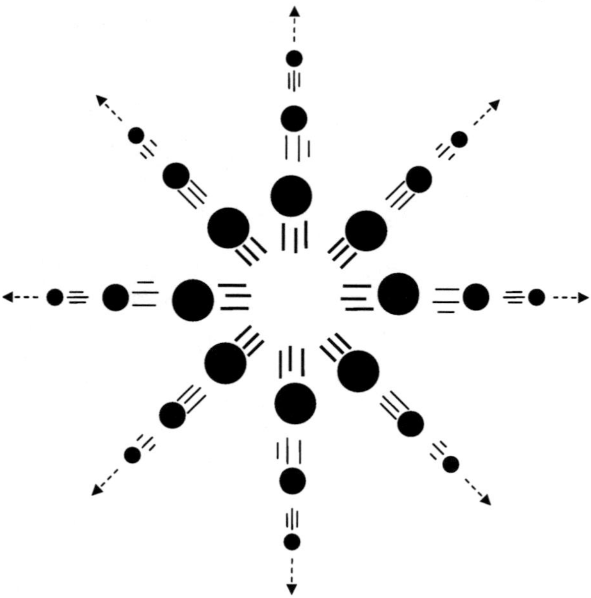

Taking all of the material objects in the universe and using their mass as rocket fuel to escape each other's gravitational pull makes the mass of the universe disappear along with all of the gravitational energy. As the rockets burn their own mass as fuel, the mass and gravitational energy become zero as the rockets approach an infinite distance from each other. See "How the Total Energy of the Universe Sums to Zero" in the Appendix for proof.

Imagine that we have a rocket ship that we will use on a journey to escape the universe. We will use the mass, m, of our rocket as fuel. Our rocket has a special engine that can convert the mass energy of our rocket into gravitational potential energy. We now set off to escape the universe. We do a series of fuel burns. Each burn consumes one-half of our rocket's mass and gets us further from the universe. Our fuel burns are shown graphically below:

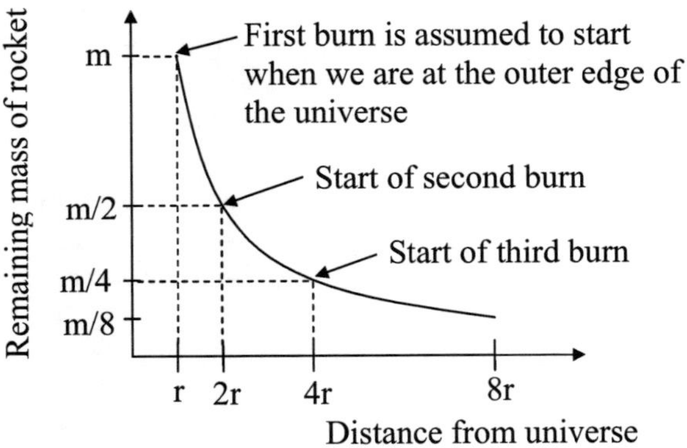

When our rocket converts half of its mass into gravitational potential energy, we will have doubled our distance from the universe. We are now twice as far from the mass of the universe and the strength of gravity has reduced by four times. At this reduced strength of gravity, the rocket can again burn one-half of its remaining mass and again double its distance from the universe. Since the force of gravity is reducing, the amount of mass needed to double our distance is also reduced for each burn. If we continue the fuel burns, our mass will reach zero along with an equal amount of negative gravitational energy. The creation process can be imagined as the reversal of this example.

Modeling the Electrical Nature of Space

The advantage of viewing the universe in terms of an electrical model lies in the fact that the electrical properties of space are well understood. Another advantage is that the electrical view may be more fundamental. According to Einstein's Theory of Relativity, the concepts of length and time depend on the motion of the observer. For example, the length of a ruler will have different values if measured by observers traveling at different relative velocities. Relativity also tells us that time itself depends on the relative velocity of the observer and that the speed of light is the same for all observers regardless of their velocities. In this way, nature seems to be telling us that the speed of light, c, is more fundamental than length or time. Since the speed of light can be represented by two more fundamental electrical parameters, as shown below, we will take nature's clue and consider the electrical model of space.

$$C = \frac{1}{\sqrt{\mu_0 \varepsilon_0}} = \text{Speed of light}$$

Where the electrical parameters are:

$\mu_0 = 4\pi E-7$ Henry/meter (The Permeability Constant is related to the magnetic or inductance properties of space.)

$\varepsilon_0 = 8.85E-12$ Farad/meter (The Permittivity Constant is related to the electric or capacitance properties of space.)

Another important advantage of viewing the universe in terms of an electrical model is that very powerful electronic simulation programs are available to analyze the behavior of these models. These same simulation programs are routinely used by industry to analyze complex electronic systems that often have millions of electrical components. These computer models of space can provide insight into the creation process. We will later show the results of such a simulation.

Our journey into the electrical nature of space will now begin. We will start by developing an electrical model of a simple universe composed of only two points.

We will assume that the creation of the universe is based on the Law of Balance:

$$A=A \text{ or } A-A=0$$

This law will now be applied to the three fundamental electrical properties of space: resistance, inductance and capacitance.

The remainder of this section may be a bit technical for some readers. Though we considered putting it in the Appendix, we decided that the average reader may find the model development steps interesting.

Resistance

If you hold the leads of an ohm meter into free space, you will measure an infinite value of resistance.

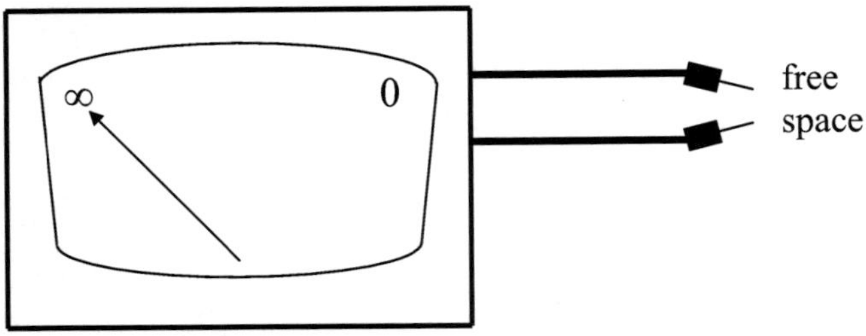

An OHM meter measures infinite resistance of free space.

Our model of a two-point universe will reflect this property of infinite resistance between the points. Recall that the Law of Balance for energy requires that the positive energy of the universe be balanced by an equal amount of negative energy, resulting in a universe with a net energy of zero (E-E=0). In a similar way, the electrical laws of the universe allow a positive resistance to be balanced by an equal negative resistance, resulting in an infinite resistance. This is accomplished by placing a positive resistance, R, in parallel with a negative resistance, -R, as shown below:

Resistance Model of a Two-Point Universe

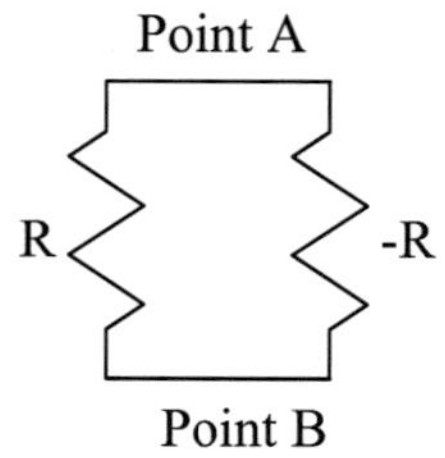

A negative resistor is not found to exist in our everyday experience, but is required by the Law of Balance. We will consider the possible existence of negative resistance as well as negative inductance and capacitance in our model formulation. If the resulting model yields reasonable results, then these negative properties need further investigation.

As an electrical engineer will tell us, the effective resistance of the circuit shown above is the product of the two resistors divided by the sum of the two resistors:

$$\text{R\_effective} = \frac{-R*R}{R-R} = \text{infinity}$$

This circuit will always yield an infinite resistance value as long as the two resistors are equal to each other. The actual resistor value does not matter.

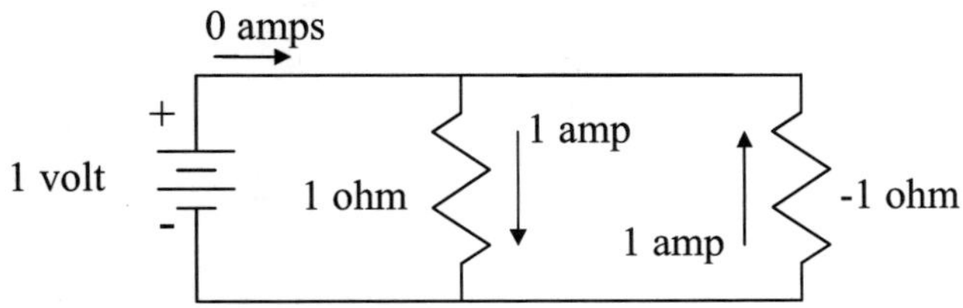

Let's consider whether a negative resistor violates the Law of Conservation of Energy. To do this, let's consider the above circuit diagram. A battery of 1 volt is applied across the resistors of 1 ohm and -1 ohm. The positive resistor on the left side will dissipate 1 watt of power and will heat up. The resistor on the right side will absorb 1 watt of power and will cool down by an equal amount. The battery will stay charged by the same amount since no current flows to or from it. The net energy of the system before applying the voltage and after applying the voltage remains the same. As long as the magnitudes of the resistors are equal, the Law of Balance is satisfied and conservation of energy is maintained.

So far, our very simple model of a two-point universe is merely a positive resistance in parallel with an equal negative resistance, resulting in infinite resistance between the points. However, this model is incomplete since the fundamental properties of inductance and capacitance have been ignored. We will next incorporate these properties of space into the model.

Inductance

Inductance is a fundamental property of space that is always present. The electrical properties of parallel inductors behave in a way similar to the properties of parallel resistors. In electrical notation, the symbol "L" is used to denote inductance, since the symbol "I" is already used to denote current. Thus, a positive inductor, L, in parallel with a negative inductor, -L, will yield an infinite value of inductance between the two points. This corresponds to the observed inductive properties of space and will thus be incorporated into our model.

Inductance Model of a Two-Point Universe

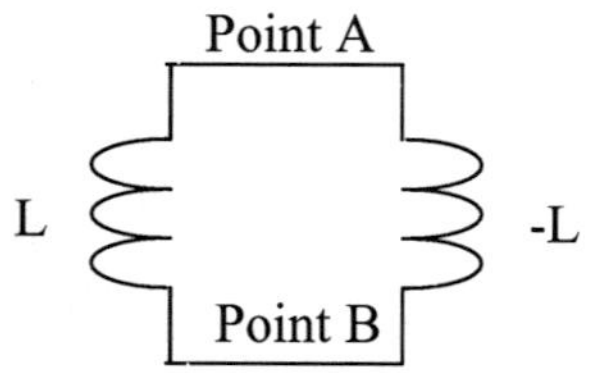

Combining the resistive properties of space with the inductive properties gives us the following LR model of a two-point universe:

Resistance-Inductance Model of a Two-Point Universe

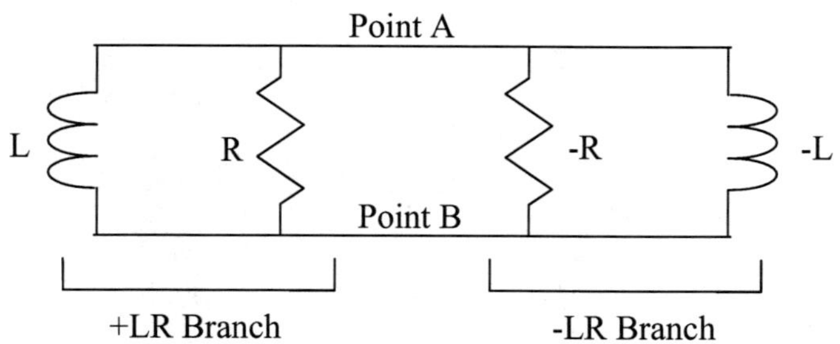

Capacitance

Like inductance, capacitance is also a fundamental property of space that is always present. Unlike resistance and inductance, parallel capacitors will be added together to get the effective electrical capacitance. In this case, applying the Law of Balance gives:

$$C-C=0$$

Thus a positive capacitor, C, in parallel with a negative capacitor, -C, will yield a total capacitance, between the two points, of zero.

Capacitance Model of a Two-Point Universe

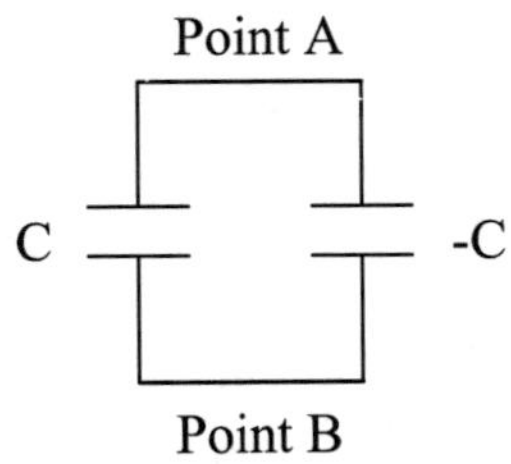

Combining our model of resistance and inductance with capacitance gives us the following model for a two-point universe:

LRC Electrical Model of a Two-Point Universe

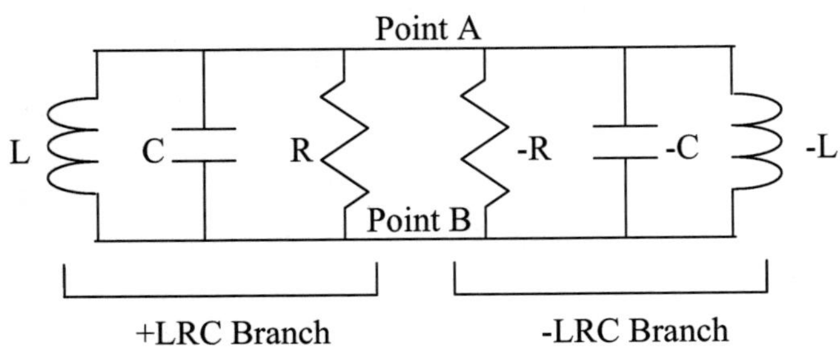

The Importance of the LC circuit

It is very significant that this simple model contains two parallel LC circuits:

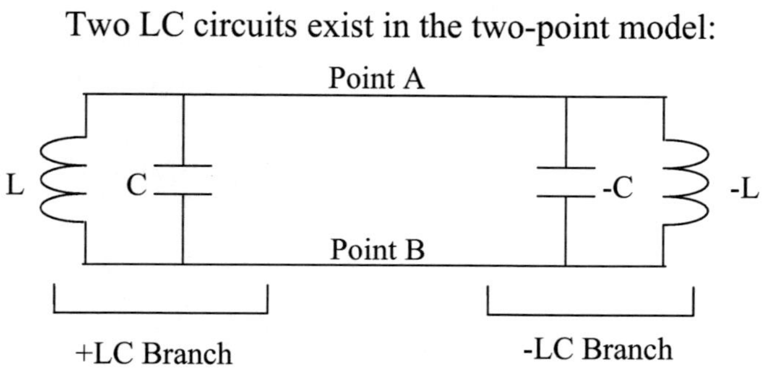

Both the +LC branch and the -LC branch are oscillating circuits capable of generating sinusoidal electromagnetic radiation, such as visible light and radio waves. As the dimensions of space shrink, so do the values of both inductance and capacitance. As the LC values shrink to zero, the generated energy heads towards infinity as the following equation reveals:

$$Energy = \frac{h}{2\pi\sqrt{LC}}$$

Where:

h = Planck's constant = 6.63×10^{-34} joules-second
L= value of inductor
C= value of capacitor

This resulting energy produces subatomic particles, including the electrons and protons, needed to form atoms. This LC circuit can also produce the observed cosmic background radiation. At higher energy levels, the more massive particles are created (protons and antiprotons). At lower energy levels, the less massive particles are created (electrons and positrons). Our model is capable of explaining the electromagnetic energy as well as the mass found in the universe.

Computer Simulation of a Self-Creating Universe

Electrical Model of a Two-Point Universe

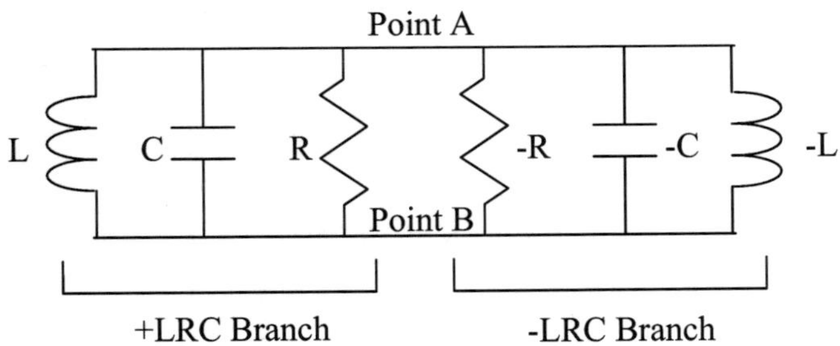

When we first simulated the two-point model, we noticed that it was very unstable. This surprised us since we have routinely simulated circuits of far greater complexity without seeing this kind of instability. Some of the numeric values calculated by the simulation program were so great that they exceeded the calculation capability of the computer. Was the model telling us that the early universe exhibited the same instability? The solution to this problem was to force the initial starting values at point A and point B to be exactly zero. When we did this, the model became stable.

We then used the simulation program to cause a very small imbalance in the circuit, corresponding to a virtual particle formation at point A. We did this by injecting a very tiny electrical charge into point A to simulate a virtual particle trigger event. What we saw next amazed us. The simulation started out with zero energy and the energy wave built up over time, becoming greater and greater. We were seeing a positive energy that was exactly balanced by a negative energy for every time point of the simulation. As the positive energy wave became larger, the negative wave became larger by exactly the same amount. The total energy of the model universe always summed to zero while the positive energy and negative energy grew over time. We could see a universe forming!

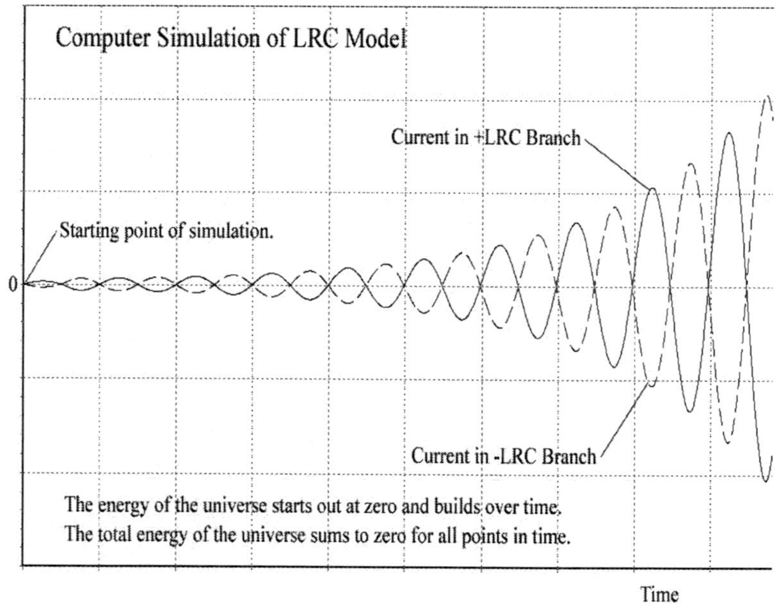

The previous picture is the actual simulation result of the model showing the currents in the LRC and -LRC branches growing over time.

We soon realized another very important result from the model simulation. We realized that a beam of charge (or current) is flowing in the positive LRC branch of the model at the same time that an equal, but opposite, beam of charge is flowing in the negative LRC branch. Each beam of charge will create a magnetic field that will cause a repulsive force on the other beam. If the beams were both flowing in the same directions, the force between the beams would be attractive. But since the beams flow in the opposite directions, the beams will repel each other. This effect is related to the well-known observation of how the current flowing in a wire will influence a nearby compass needle. It can be shown that the force of repulsion, Fr, experienced by each of the beams is:

$$Fr = \frac{\mu_0 d\, I^2}{2\pi x}$$

Where:

Fr = force of repulsion on each beam of charge
d = distance between point A and point B
I = beam current
x = separation between the beams

As the beam currents build up over time, we find that the repulsive force between the beams increases as the square of the beam current. Thus, an enormous repulsive force builds up over time, forcing the beams apart and expanding the universe. As the current beams separate from each other, the two point universe goes from one dimension into a two dimensional ring of oscillating current. The laws of electro-magnetism require this ring of current to produce a magnetic field at a 90 degree angle to the surface of this ring. This causes the third spatial dimension to form. Thus, we can see how the simple two point universe can transition from one dimension into two dimensions and finally into three dimensions. This oscillating ring of current along with its magnetic field produces a tiny magnet with a north pole and a south pole. Could space be an array of these tiny magnets?

These results exceeded our expectations. By applying the Law of Balance to the electrical nature of a two-point universe, we derived an electrical model that helps to explain:

1. How the energy of the universe can come into existence from nothing without violating the Law of Conservation of Energy.
2. How electromagnetic radiation is produced in the early universe, resulting in the particles that form atoms as well as the observed cosmic background radiation.
3. How the enormous repulsive force between LRC branches in the two-point universe causes space to be pushed apart, resulting in a possible Big Bang explosion.

Star and Galaxy Creation

The early universe produced mostly dark matter along with hydrogen atoms and a lesser amount of helium atoms. The dark matter is invisible, but its presence can be deduced from its gravitational influence on the visible matter. During the early expansion of the universe, any local random increase in the density of matter would have increased the gravitational attraction of this local region. This causes other nearby hydrogen atoms to be attracted to this region, resulting in local hydrogen clumping. As the universe continues to expand, these local hydrogen clumps will expand at a slower rate due to the increased inward attraction of gravity. This will further accelerate the clumping of large hydrogen clouds. This snowball effect sets the seeds for galaxy and planetary formation.

These large hydrogen clouds will continue their inward attraction until the outward pressure between the atoms can balance the compression force of gravity. As these early clouds shrink inward, their rate of rotation will increase. This is similar to the increased spin that we see when an ice-skater pulls in her arms during a spin. When we view the universe, we see the results of this spin in the rotation of spiral galaxies, in the rotation of planets around a central star, and in the rotation of the planets about their own axis. The next photograph is an example of such a rotating galaxy.

Photograph of a rotating spiral galaxy.

Photograph of a region of hydrogen and dust where stars are forming.

A Rotating Hydrogen Cloud Forming a Galaxy and Planetary System

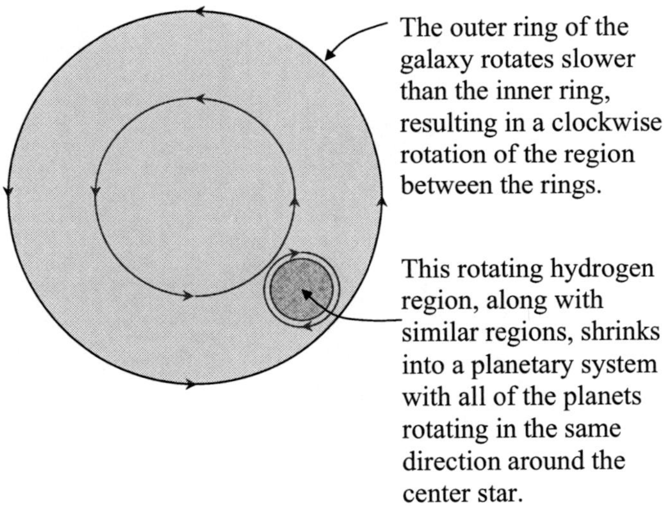

The above picture shows a counterclockwise-rotating hydrogen cloud that is in the process of forming a galaxy. Note that the center region rotates at a faster rate around the galaxy center than does the region farther from the center. This is similar to how a bathtub drains water. The water nearest to the drain hole rotates the fastest. This difference in the rate of galactic rotation causes local rotation of smaller regions, resulting in planetary systems. In general, all of the planets would be expected to rotate in the same direction around the center star, as does our own planetary system. Exceptions have been observed, however, when the planetary system was formed from multiple clouds that rotated in different directions.

Under the force of gravity, a rotating cloud of hydrogen gas will spread out into a thin disk. A similar example is found in the thin rings of Saturn, where an ancient moon was ripped apart by Saturn's strong tidal forces.

The rings of Saturn are believed to have been formed from a moon that got too close and was ripped apart by Saturn's tidal forces. Note how this once-spherical moon now forms a thin rotating disk. The strong tidal forces prevented the reformation of a moon. As will be shown in the next picture, a similar thin disk is formed during planetary formation.

In the picture below, notice how the rotating hydrogen cloud spreads out into a thin disk, similar to the rings of Saturn.

Example of Planetary System Formation

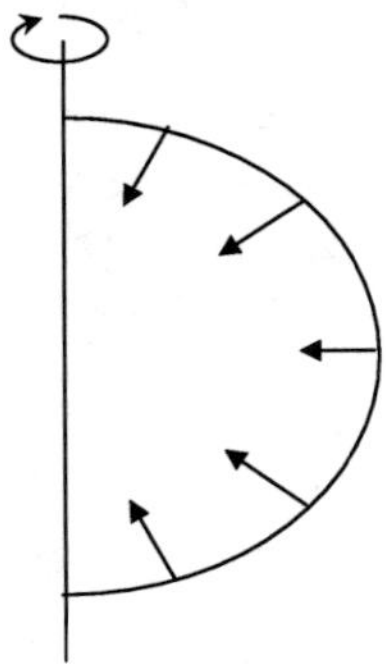

Shown is the right half of a slightly rotating hydrogen cloud that is shrinking under the pull of gravity. Note that the direction of the pull is toward the center of the rotating mass. As the cloud shrinks, the rate of rotation increases, resulting in the formation of the rotating disk shown below.

The center region of this rotating disk will form a star. Under the pull of gravity, the outer regions will form planets. The inner region rotates faster than the outer region, resulting in local rotations within the disk. This explains moon formation and the spin of the planets about their own axis.

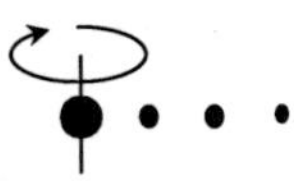

Resulting planetary system

How the Stars Create the Heavy Elements Needed for Life

The Big Bang explosion produced a visible universe of approximately 74 percent hydrogen and 24 percent helium, with some trace amounts of other elements. It is from these early ingredients that large regions of space condensed under the attractive force of gravity to create galaxies and planetary systems. At the core of these planetary systems are the stars that take these early ingredients and create the heavier elements needed for life.

One of the greatest triumphs of modern astronomy is that we know the composition of the universe. It is composed of the same type of atoms that are found here on earth. The light that we receive from distant objects reveals their chemical composition. Each atom and molecule emits and absorbs light at specific frequencies. For example, the specific frequencies of the light emitted by a hydrogen atom are unique and different from the frequency spectrum of any other element or molecule. Each element and molecule has its own unique frequency fingerprint. Whether observed in distant regions of space or on earth, these fingerprints are identical. The next picture shows the unique frequency fingerprints for some common elements.

The unique frequency fingerprints of some common elements

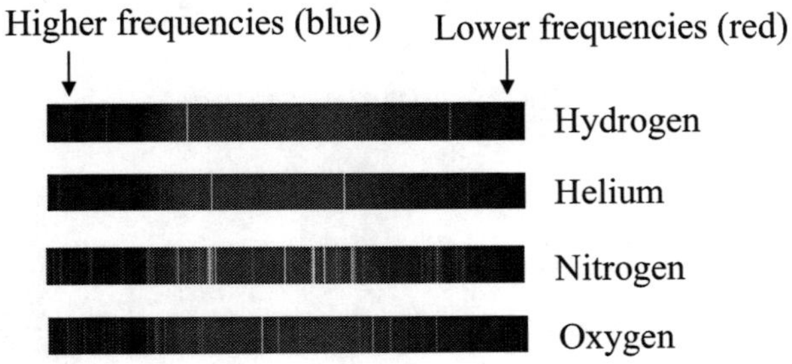

The atom can be viewed as a small solar system in which the electrons, acting like planets, orbit the more massive central core that contains the protons and neutrons. This analogy is an oversimplification but is helpful to convey the basic concept. The lines in the spectrum of the atoms are a result of the electrons changing orbit locations. This can happen when atoms interact with nearby atoms or when a pulse of light strikes the atom.

When an electron transitions from a higher orbit to a lower orbit, it will give up energy. This reduction in energy produces a pulse of light. The atom can also absorb light when a pulse of light kicks an electron into a higher orbit. An atom will only absorb or emit light of certain distinct frequencies that depend on the electron orbital configuration. Since each orbital configuration is unique, the light fingerprint of each atom is unique.

How the Atom Emits Light

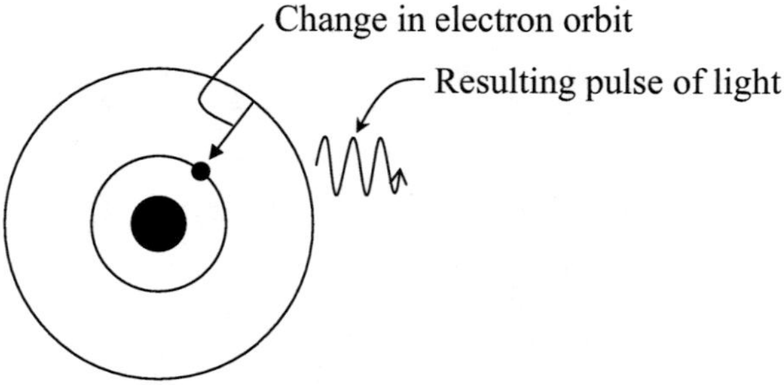

The atom emits a pulse of light when an electron gives up energy by going from a higher orbit to a lower orbit. The atom can also absorb light when a pulse of light kicks an electron into a higher orbit.

By studying the frequency components of the light, we can determine the chemical makeup of distant objects. By observing the frequency components of the light from the sun, scientists found the element helium on the sun before it was discovered on earth. In 1952, the astronomical community was surprised when the light from Red Giant stars was found to be rich in the element technetium. Since the longest-existing isotope of technetium has a half-life of only 4 million years and the universe has existed for about 14,000 million years, this is considered proof that the element technetium is being produced by stars.

To understand how the stars create the heavy elements, let's first consider the nature of temperature. A star is composed mostly of hydrogen and helium gases. As any gas is compressed, it will heat up. For example, when the piston in an automobile engine compresses the gas in the cylinder, the gas will heat up. In this case, the gas molecules are pushed closer together in a more confined space. The collisions between the gas molecules are now more numerous and at a higher speed. This faster vibration of the gas molecules results in a temperature increase. This same temperature increase occurs in stars when the gases are compressed by gravity. The opposite is also true. As any gas is allowed to expand, it will cool down. We feel the cool mist when a compressed gas is released from a spray can.

What is Temperature?

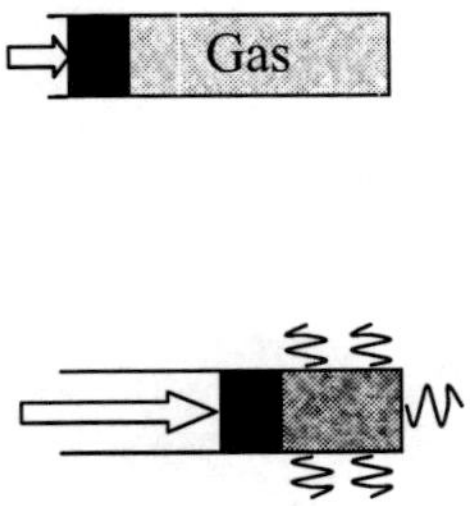

To the left is shown a gas-filled cylinder being compressed by a piston. As the piston moves to the right, the gas molecules and atoms collide with each other at a faster and more energetic rate. This increase in thermal vibration results in a temperature increase. The further the piston is pushed to the right, the hotter the gas becomes.

In a similar way, the gas in a forming star is compressed by the force of gravity, causing the star to heat up. As the star compresses and gets hotter, the outward pressure of the compressed hydrogen and helium in the star's core will eventually balance the inward pressure from gravity. When this balance is achieved, the star becomes stable.

Star Pressure and Temperature

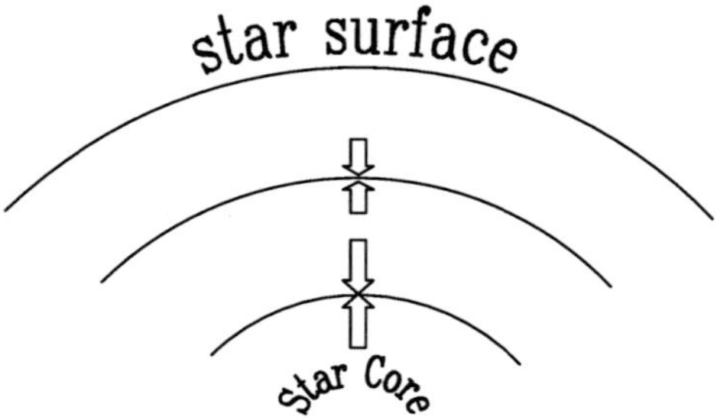

The inward pressure of gravity and the balancing outward gas pressure exerted by the compressed core are shown for two different depths. Since each layer must support the weight of the layers above, the pressure, temperature, and density all increase at depths closer to the star core. Hydrogen-to-helium conversion begins at the star's core.

The Building Blocks of Matter

(+) The proton has a positive charge. The nucleus of the hydrogen atom is just a single proton. Under the extreme pressure and temperature in a star, the proton can combine with an electron to create a neutron.

(n) The neutron has no charge. It helps to stabilize the nucleus of the atom. The mass of the neutron is slightly more than that of the proton.

⊖ The electron has a negative charge that exactly balances the positive charge of the proton. Electrons exist in orbits around the nucleus. The electron is about 2000 times less massive than the proton.

How a Star Creates Helium from Hydrogen and Produces Energy

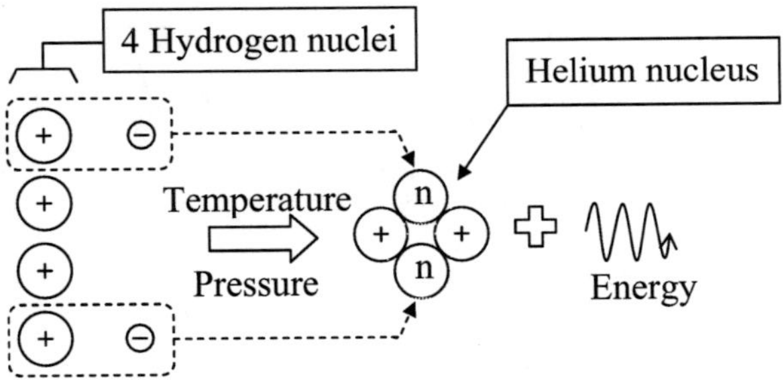

Four hydrogen nuclei combine with two electrons to produce one helium nucleus plus solar radiant energy. The mass of the resulting helium nucleus is less than the mass of the starting hydrogen nuclei and electrons. This difference in mass is converted to energy according to $E=mc^2$. In the case of the sun, some of this energy is radiated to earth to allow and sustain life.

Helium Next Burns into Carbon:

As the star runs out of hydrogen to burn, the temperature and pressure inside the star's core will start to decrease. This lack of internal core pressure will allow gravity to further compress the star. This added compression causes the star core to heat back up.

The compression continues until the temperature of the star heats up enough for helium to start burning into carbon. When carbon is created from helium, energy is liberated. This increase in liberated energy will cause the star's core to heat up until the internal star's outward pressure balances the inward pressure of gravity.

Carbon Plus Helium Burns to Make Oxygen:

Helium keeps burning to carbon until the helium fuel is nearly exhausted. Again, the internal core pressure of the star drops, allowing gravity to further compress the star until a temperature is reached at which carbon can fuse with remaining helium to create oxygen.

The Burning Process Continues Until Iron is Made:

This process continues to create such elements as silicon, sulfur, and calcium until, finally, the element iron is created. As we know, iron is a very heavy and dense metal. The very dense state of the iron atom prevents the iron from being compressed further. As a result, the creation of elements heavier than iron are delayed. At this point, the solar furnace starts to shut down.

As the elements evolve, complex elements are made from less complex elements. The cosmic abundance of the elements decreases at the higher atomic weights. Thus, the cosmic abundance of the later-created elements declines. Light elements such as hydrogen and oxygen are very common and heavy elements such as gold and platinum are very rare. The chemical evolution of some common elements is shown in the following chart:

The Stellar Creation of Some Common Elements

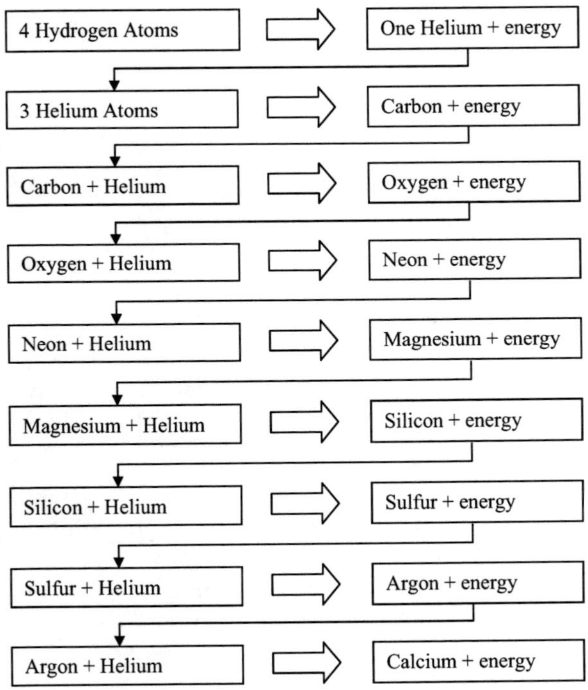

As a massive star ages, rings of consecutive elements build up, with the lighter elements at the surface and the heavier elements at the core.

Element Buildup in a Massive Star

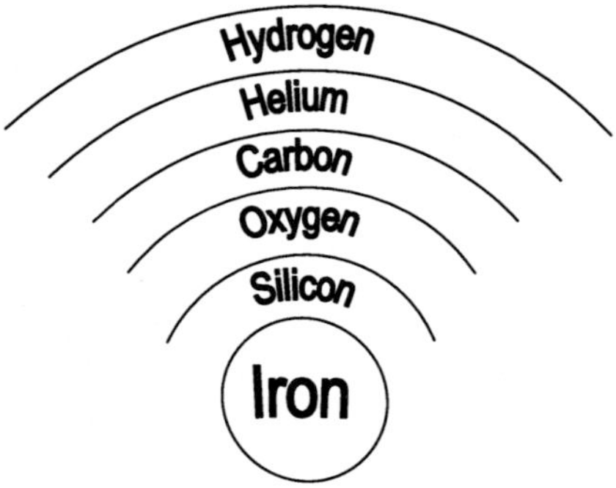

Initially, the star burns hydrogen into helium. The helium core then burns to carbon and oxygen. At each stage, the core is burning a more complex element. This process continues until iron is reached. At this point, the normal burning process stops. The star will now cast off much of its mass, allowing the next generation of stars to be composed of more complex elements. In this way, matter evolves from simple to complex.

A normal star will burn hydrogen for most of its life cycle. In general, consecutive elements will burn faster and at higher temperatures. Near the end of the star's life, it will greatly increase in radius and become what is called a Red Giant—or, in the case of a very massive star, it will become a Red Super Giant. Our sun will someday become a Red Giant and then a Planetary Nebula, casting its mass into the cosmos for the next generation of stars. A star of about nine times the mass of the sun or larger will become a Red Super Giant and end its life in a supernova explosion. The supernova explosion will form elements heavier than iron and expel its mass into the universe to form new stars.

The Life Cycle of a Star:

The First Phase in the Life Cycle of a Star

This is the stable period of a star. During this phase a typical star will burn hydrogen to helium. Our sun will spend about 10 billion years in this phase. A more massive star will generate more heat, burn its hydrogen fuel faster, and have a shorter life span.

The Star Expands in Size During the Second Phase

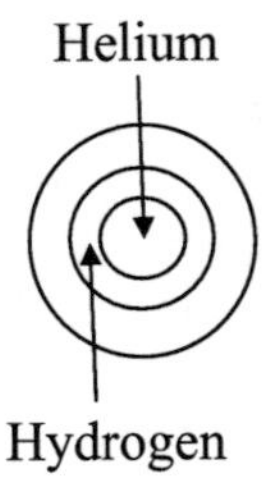

Near the end of a star's life cycle, the central core will become enriched with helium and the surrounding ring will still burn hydrogen. As the central helium core decreases in temperature, its pressure will decrease. This will cause gravity to compress the helium core. As this happens, the surrounding hydrogen ring will heat up and burn faster and the star's luminosity will greatly increase. This hot-burning hydrogen ring will push away the outer regions causing the star to expand into a Red Giant or, in the case of a very massive star, into a Red Super Giant.

The Third Phase Recycles the Star's Mass

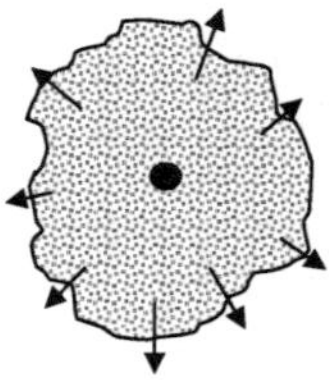

As the outer layers of the Red Giant are pushed off into space by the hot burning core, a Planetary Nebula is formed. A more massive star will result in a Supernova explosion. In either case, much of the star's mass is cast off into space to be recycled into the next generation of stars. In this way, the matter evolves.

A Planetary Nebula Seeding the Universe for the Next Generation of Stars

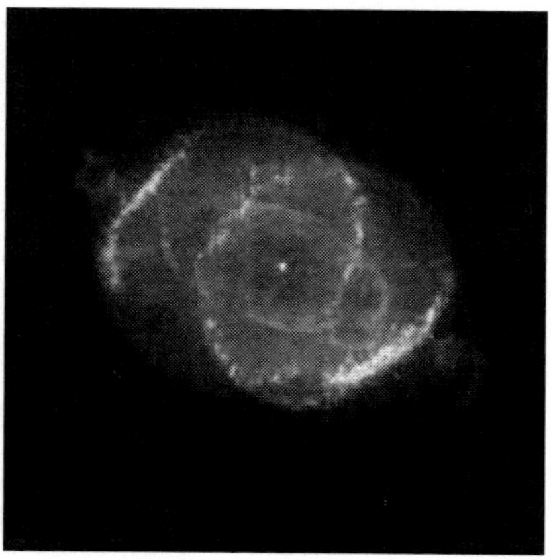

Above is a photograph of a Planetary Nebula. In about 5 billion years, our sun will enter the Red Giant phase. The sun will next form a similar Planetary Nebula casting off much of its mass. In this phase, the core temperature becomes so hot that the outer layers are pushed away.

Seeding the Universe by a Supernova Star Explosion

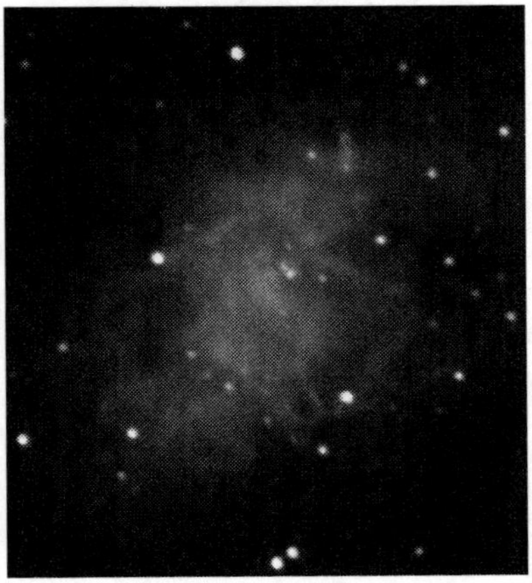

Above is a picture of the Crab Nebula Supernova explosion that occurred in the year 1054. A star of about 9 solar masses or more will end its life in a supernova explosion. During the explosion, enough energy is released to create elements heavier than iron. Rare metals such as gold, silver and platinum were created in a supernova explosion.

The stars take the simplest elements that were formed in the Big Bang explosion and create more complex elements. Each new generation of stars has the elements that were formed by older generations. Stars provide the mechanism for the universe to chemically evolve. In the early universe, the chemistry was not sufficiently evolved for human life. It may have taken several generations of stars to create the elements necessary for the formation of life.

The Start of Life

Analysis of ancient rocks reveals that the earth's early atmosphere contained little oxygen. The abundance of oxygen in today's atmosphere is a result of the balance between plant and animal life. The plants create atmospheric oxygen that the animals breathe in and convert to carbon dioxide. The plants then take in the carbon dioxide and produce oxygen. The earth's primitive atmosphere is believed to have contained hydrogen, helium, methane, ammonia and water—similar to the atmospheres found on Jupiter and Saturn, where life never gained a foothold.

Atmospheric Content of Jupiter

(By volume)

Hydrogen (H_2)	89%
Helium (He)	10.2%
Methane (CH_4)	0.3%
Ammonia (NH_3)	0.026%
Water (H_2O)	0.0004%

Atmospheric Content of Saturn

(By volume)

Hydrogen (H_2)	96%
Helium (He)	3%
Methane (CH_4)	0.4%
Ammonia (NH_3)	0.01%
Water (H_2O)	0.2%

In 1953, Stanley Miller and Harold Urey at the University of Chicago conducted a famous experiment. They mixed the ingredients of the earth's early atmosphere in a sealed glass apparatus and then reproduced the earth's early environment. The experiment used a mixture of hydrogen, methane, ammonia and water. Helium was excluded since it is chemically inactive. These ingredients were sealed in a glass tube apparatus to form a circulating loop similar to the evaporation-rain cycle on earth.

The apparatus contained a primary bowl to simulate the earth's primitive ocean. The contents of this bowl were heated to cause evaporation into a second chamber that simulated the earth's early atmosphere. The contents of this second chamber underwent electrical discharge to simulate lightning. The contents of the second atmosphere chamber were then cooled to form rain. The rain then returned to the primary ocean bowl where the cycle repeated. A diagram of the experiment is shown in the next figure.

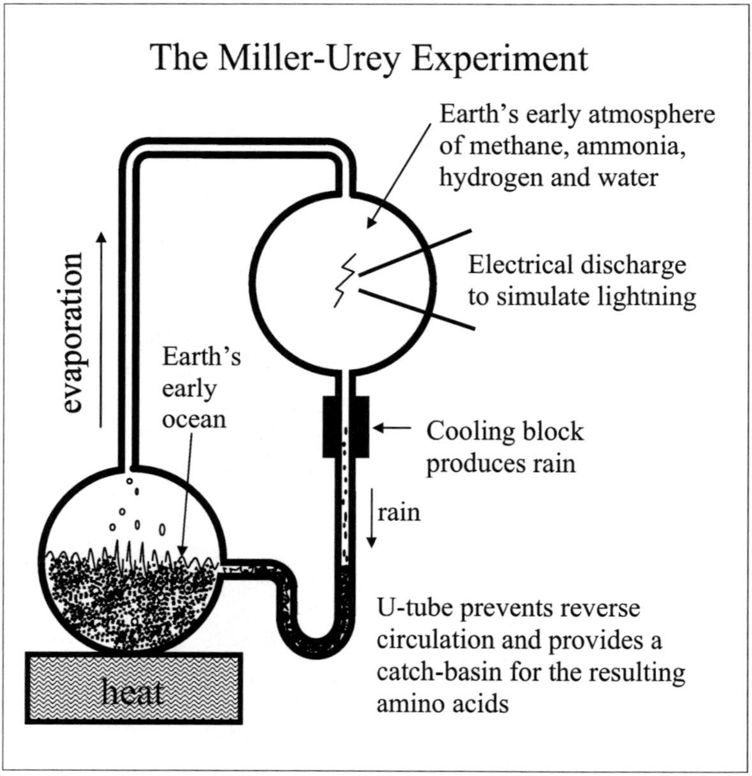

After the experiment ran for one day, the water changed to a pink color. After one week, the solution had turned to a deep red along with yellow organic compounds. The resulting materials were then analyzed and found to contain amino acids, the basic building blocks of life. Other variations of this experiment resulted in the creation of all of the 20 primary amino acids required for life. Amino acids link with each other to form linear protein chains.

Amino Acids Link with Each Other to Form Proteins

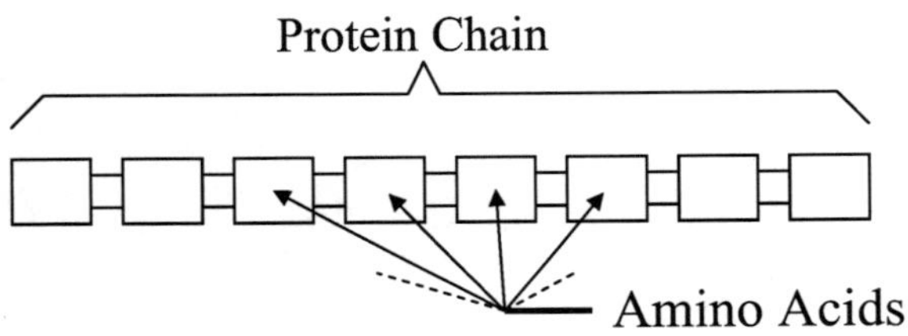

Some parts of the human body made from amino acids include:

- Skin
- Muscles
- Hair and nails
- Tendons and ligaments
- Eye lens
- Blood vessels

In addition to occurring on earth, amino acids have also been found in the tails of comets, in Red Giant stars and in meteorites.

For amino acids to link up and form complex proteins, a protective environment is needed. Just as we live in houses to protect us from the wind and rain, the early proteins needed protection from a hostile environment in order to evolve into more complex molecules. In the 1950s and 1960s, the biochemist Sidney Fox, working at the University of Miami, found such a protective environment. He found that mixtures of amino acids could easily form small spherical shells with diameters of about 2 microns. These cell-like spheres could reproduce in a cell-like manner by applying pressure that caused a micro-bubble to split into two micro-bubbles.

Cell-Like Spherical Bubble Reproduction

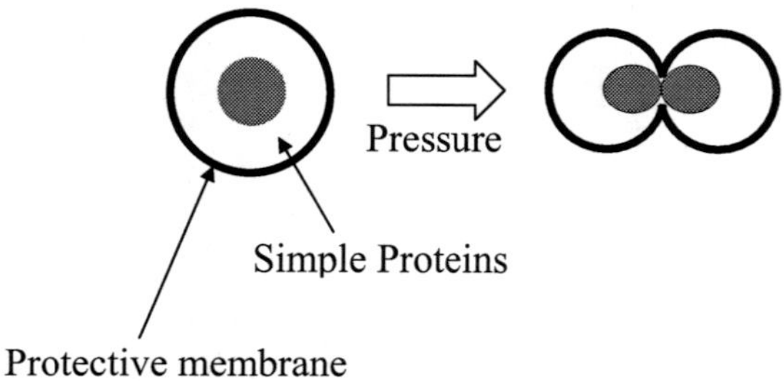

Depicted above is a micro-sphere as a stepping stone between organic molecules and primitive cellular life. Professor Sidney Fox observed cell-like reproduction when pressure was applied to the microscope slide cover glass.

The first cells on earth had to have a way of getting nutrients through their membranes in order to grow. Scientists have found that fat molecules can be used to construct simple cell membranes that allow nutrients in. Fat molecules, believed to have existed prior to the formation of life, have been found to spontaneously ball up into tiny spheres in the presence of saltwater.

Research led by the Howard Hughes Medical Institute has shown that small bubble cells of fatty acids filled with organic material could compete in the "survival of the fittest" competition by drawing material from other cells. The membrane walls of these small bubble cells were simple compared to living cells. Researchers filled fatty acid bubbles with either an organic sugar solution or with a sugarless solution. The bubble cells with the sugar solution developed more internal cell pressure and expanded. When the researcher combined both types of cells, they found that the cells containing the sugar had greater membrane pressure and were able to grow by drawing membrane material from nearby cells that did not contain sugar. These cells were feeding on each other without being alive.

The Miller-Urey Experiment showed that the earth's ancient sea floor and tide pools likely served as collection basins for the formation of early organic matter, including amino acids and simple protein chains. It is also found that this organic matter can form tiny spherical bubbles, or micro-spheres, to house the early amino acids and protein links. The earth's heat from sea floor thermal vents and from solar energy heated up tide pools, causing these bubbles to expand and rise to the surface and release the captured ingredients into the atmosphere. UV radiation along with lightning, as the Miller-Urey Experiment showed, can chemically transform these ingredients. These newly transformed ingredients sank back to the ocean floor providing nutrients for the next generation of spherical bubbles. This cycle of (1) bubble formation at the ocean bottom, (2) rise to the surface, (3) release into the atmosphere, (4) chemical transformation and (5) sinking back to the ocean bottom would be expected to repeat itself throughout millions of years.

Thus, an evolutionary cycle is set into motion where the stable micro-spheres will survive, allowing their internal contents to further evolve. The unstable spheres will burst and become the food for the stable spheres. This spherical bubble cycle yields stable structures that could eventually evolve into simple reproducing cells. Thus, the early chemical evolution and the later evolution of life may be a very natural process.

Earth's Bubble Cycle

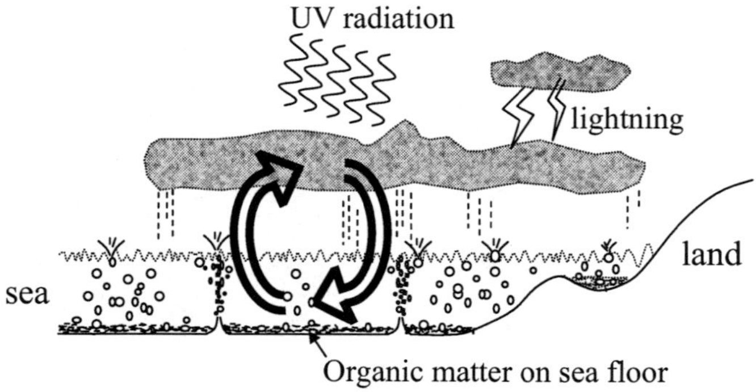

Cell-like spherical bubbles form at the sea floor from decaying organic matter and rise to the surface releasing ingredients into the atmosphere. UV radiation and lightning alter their chemical structure returning the ingredients back to the sea as rain. This cycle is repeats for millions of years. The stable micro-bubbles evolve into primitive cellular life, drawing nutrients from the surroundings.

There exists a rough correspondence between blood and sea water content. This provides a hint that primitive life incorporated the surrounding environment. In the early 1900s, Rene Quinton replaced all of the blood content of a dog with diluted sea water. The dog re-generated its blood in about four days and went on to live a happy life. In emergency transfusions, when blood is not available, diluted sea water has been used in blood transfusions. This happened during World War II. Navy doctors first diluted sea water with distilled water. Early sea water may have been less salty since the ocean's salt content is increased by land erosion.

What about the Eye?

It is often claimed that a system as complex as the human eye could not have evolved and that such a complex mechanism is proof of an intelligent designer, just as a fine-jeweled watch is proof of a watchmaker.

Let's consider how the eye might have evolved through a series of small evolutionary steps. Let's start with a simple, single-celled organism that can swim about, using its tail as propulsion. We will assume that this simple organism has a clear, jellyfish-like body. This creature is competing with similar organisms to feed on algae-like material found in warm tide pools near the sea coast. Our creature finds its way around by sensing objects that it has bumped into. Our creature is shown below.

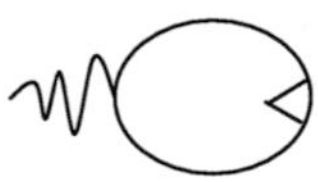

Hypothetical one-celled creature that has no eyes.

Let's assume that one of our creatures has undergone a mutation through which it has developed a dark mole or freckle on its body.

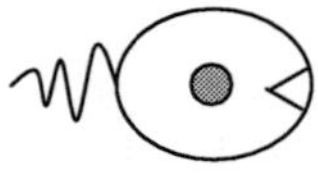

Our creature has undergone a mutation by which it developed a brown spot on its body, similar to a freckle.

As our mutated creature swims near the surface of the sea, the sunlight heats the freckle more than the rest of the creature's body, because the darker material absorbs more heat. Swimming in shallow tide pools close to land, the creature senses the strong solar heat. Our creature realizes that when the heat is more intense, it is easier to find food. This provides a big advantage over its competitors and our creature grows and reproduces many times, passing on the dark spot to its descendants. The descendants soon edge out the creatures without the spot.

Some of the creatures have darker spots that receive more solar heat, making the richer food sources easier to find. These creatures will out-feed the ones with lighter spots and pass on the darker spots to the next generation. This spot will eventually become black, like the cavity inside the human eye.

After many generations, a second mutation occurs. This mutation causes a small pocket of clear, jelly-like material to form on the top of the black spot. This new material could be genetically identical to the clear jelly material of the rest of the body. It acts like a lens and helps to focus the light, generating a more intense heat. This creature now has an even easier time finding the warm, shallow tide pools where the food is plentiful.

Clear jelly sac acting as a primitive lens

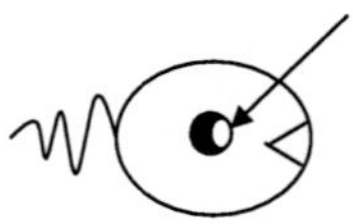

The second mutation creates a clear jelly-like region on the black spot. This focuses the sunlight providing a more intense heat on the black spot.

Over many generations of evolution, this crude mechanism develops more nerve connections and the lens becomes more refined. Small muscles develop that can change the shape of the lens and allow controlled focusing of the light. At each stage, the new generation has an advantage in finding food and is able to survive and reproduce. A third important mutation occurs that causes the genetic code to produce a second similar eye. This offers better vision and a backup eye. Ultimately, a human-like eye evolves.

Evidence that the eye evolved and was not designed lies in the fact that the human eye is a rather poor design. Besides allowing macular degeneration, cataracts, retinal detachment, astigmatism, presbyopia and glaucoma, the eye has some design features that would make any engineer cringe. For example, the blood vessels are actually between the sensors and the eye lens, causing vision distortion. A sensible design would have put the blood vessels at the bottom of the sensors, away from the light source.

Humans have been able to develop electronic eyes, based on the silicon atom, that are far superior to the human eye. For example, the electronic eye on a space satellite can spot an object much smaller than a golf ball on the earth's surface. The human eye is far from having this ability. The electronic eye has a zoom feature that is missing from the human eye. It also has the ability to view longer wavelengths and can see through clouds. It has a much faster response to changes in movement and will last much longer than the human eye. Since simple humans can design an electronic eye that is far superior in performance to their own eyes, it is very unlikely that a superior intelligence designed the inferior human eye.

Putting it All Together

If we set out to design our own universe, we would soon learn that nature has fundamental restrictions on what we can create. We would find that any circular object must have a ratio of circumference to diameter with a value of π (3.14159…). Any spherical object must also have a definite relationship for its volume and surface area that also depends on the constant π. We have no other choice if we want to have spherical objects in our universe.

If we want our universe to have planets and stars that are cubes rather than spheres, then we could not use the inverse square law of gravity that clearly depends on the properties of the sphere and of π. After attempting to design a universe with cubic planets and stars, we would soon throw our hands up in the air in frustration, and instead use the simple inverse square law of force. We would quickly discover that we have little choice in what we can design.

Nature transfers energy from place to place using sinusoidal energy waves that depend on the fundamental properties of the circle. If we decided that we wanted to design a universe that transferred energy in the form of square waves rather then sine waves, then we would soon learn that we have gotten ourselves into a really big mess. The energy required for the rapid edge transitions of the square wave would require near infinite energy and we would soon opt for the sinusoidal method. We would soon end up with a design that would be very similar, if not identical, to what we have.

If we wanted to create our universe out of nothing, this would be the most difficult task of all, since the state of nothing is unstable. It is also a violation of the laws of quantum mechanics. We would need to be very careful in how we handled this nothing, since it is like a bomb ready to go off at the slightest disturbance. If we could actually create the state of nothing, then we would expect the universe to occur naturally without any need for further design on our part. The real task would be how to stop a universe of nothing from self-creating the universe that we see when we look out into space.

As we have seen, the beginning and evolution of the universe is the result of very simple principles. Basic mathematics teaches us that anything divided by itself is equal to 1. For example, 7 divided by 7 is 1. Pick up your calculator and see what you get if you divide zero by zero. Instead of getting one, you get an error. This is because zero divided by zero can be anything—including our universe. As previously shown, all of the key energy laws contain a zero divided by zero for a universe of nothing. A universe of only two objects of zero mass, separated from each other by zero distance, can have any value of energy. The energy law for gravity in the case of "nothing" is:

$$Energy = G\frac{m_1 m_2}{d} = \frac{0}{0} = anything$$

The "anything" resulting from this law must be a stable universe. Any other solution to the equation will not remain.

It takes very little to trigger an unstable "nothing" into existence. An all-knowing super being is not needed for this task. This would be like using an atomic bomb to move a grain of sand. The slightest vacuum fluctuation will start our universe on its way.

It is not enough that the laws of physics allow for the self-creation of the universe. We must understand the basic principle that allows a universe of something to be created without violation of the Law of Conservation of Energy.

The simplest truth of all is the Law of Balance. This states what we consider to be self-evident, that something is equal to itself. This law is stated mathematically as:

$$A=A$$

Doing simple algebra, this law becomes:

$$A - A = 0$$

This is the simplest of all principles, where the first "A" term represents the total mass energy of the universe and the second term, "-A," represents the total balancing negative gravitational energy. The total energy of the universe always sums to zero so that the Law of Conservation of Energy is never violated. Could it be that the root cause of the universe is that something is equal to itself and that "nothing" divided by "nothing" is equal to "anything"? Could it really be that simple? Why not? If it is really this simple, then we should ask a deeper question: Where did these simple principles come from?

Why can nature's zero divided by zero take on any value? Is it just something that cannot be any other way, like the constant π? When we consider "A=A", is any other choice possible? Could the equation "A=A" be a statement that contains the Law of Conservation of Energy?

We can imagine that before the Big Bang, there was a static universe of nothing. Since time is a measure that depends on the relative motion of two or more objects, a universe of nothing has no time. It makes no sense to talk about what happened before the Big Bang since "before" did not exist. Somehow, things went from a state of nothing to the universe that we see today.

The Heisenberg Uncertainty Principle of Quantum Physics provides insight into the mechanism that triggered the universe into existence. One of the outcomes of this well-established law of nature is that it is impossible for any particular point in space to exist for long without energy. A perfect vacuum will not last since the state of "nothing" is unstable. Nature outlaws non-existence. Virtual particles are always popping into existence and going out of existence. This has been well verified by physical measurements of the Lamb Shift and the Casimir Effect.

We might want to consider why the Heisenberg Uncertainty Principle exists in the first place. Did some super being design this principle to trigger the universe into existence? The existence of the Heisenberg Uncertainty Principle is similar to the existence of the constant π. If the circle does not exist, then does π exist? As soon as a circle exists, the ratio of its circumference to diameter also exists. As π is an integral part of a circle, so is the Uncertainty Principle an integral part of the universe.

The electrical model of a two-point universe, which we previously explored, demands the existence of a universe. By inserting a virtual particle into the model, a universe oscillates into existence. This also makes intuitive sense: as soon as the first particle pops into existence, an enormous pressure difference exists between the particle boundary and the "nothing" that exists outside of the particle. This enormous pressure difference creates the Big Bang explosion.

As the Big Bang creates the universe, mathematical analysis presented in the Appendix shows that the positive energy in the form of mass is exactly balanced by the negative energy in the gravitational field of space. This results in a universe where the total energy is zero at every point in its creation. This positive versus negative energy balance also agrees with the electrical model simulation results.

There were enormous levels of temperature and pressure as the explosion got under way. This phase resulted in the creation of hydrogen and helium from the initial electromagnetic radiation. As the universe expanded and cooled, the hydrogen and helium production quickly stopped, since the energy of the radiation dropped below the mass energy of the hydrogen and helium atoms.

The hydrogen and helium atoms created in the early universe condense into clouds under the attractive force of gravity and then condense into galaxies and into solar systems. The stars convert the early hydrogen into the heavier elements and then cast the newly created elements off into space. These elements are then recycled into the next generation of stars. After several generations of stars, enough heavy elements have evolved to allow the existence of life.

The laws of probability, derived and presented in the Appendix, show that, given enough time, even the most unlikely event will occur. Given the enormous expanse of the earth's oceans and the early conditions in the earth's atmosphere, simple self-replicating molecules developed. The evidence of the earth's biological evolution has been well established. The universe is an evolution from simple to complex elements and then from simple to complex life. As with the formation of an intricate snowflake from simple water molecules, nature has the ability to take the simple and create the complex.

Appendix

A Brief Overview of Calculus

Knowledge of algebra is sufficient for the design a fine home, while calculus is needed to design a space craft and visit the stars. Two fundamental concepts of calculus are differentiation and integration. Differentiation is the process of finding a derivative. A derivative is the rate of change of something. Just think of the word "derivative" as meaning "change." For example, acceleration is the time rate of change of velocity, or we can say that acceleration is the time derivative of velocity. Velocity is the time rate of change of position, so we can say that velocity is the time derivative of position. The derivative is an instantaneous rate of change rather than a rate of change averaged over time.

The derivative can also be considered as the slope of a graph at a given point on the graph. The slope at a point on a velocity-versus-time graph is the acceleration at that point.

We will show that the fundamental concept of integration is just the reverse of the differentiation process. Integration can be considered the process of finding the area under a graphed curve. Differentiation can be considered the process of finding the slope of a graphed curve. These concepts are next illustrated.

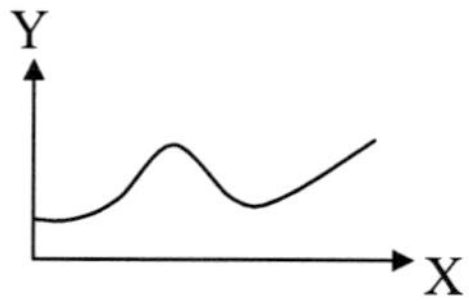

This is a graph of an equation showing a relationship between Y and X.

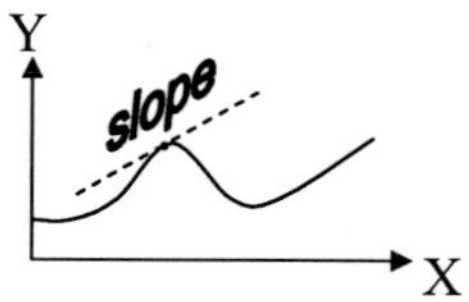

Differentiation is the process used by calculus to find the slope at a particular point on the graphed equation.

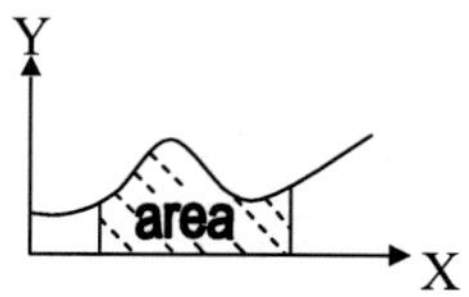

Integration is the process used by calculus to find the area under a particular part of the graphed equation.

The essence of calculus is to examine at the very minute level and in doing so, small error terms will go away, enabling an exact result. When the distance, x, is examined by calculus at the very minute level, this minute distance is called "dx". This distance is as close to zero as possible without becoming zero.

A few examples will help to make this clear.

Shown next is the graph of the function $y=x^2$. We will use calculus to find the derivative (slope) at any point on the curve.

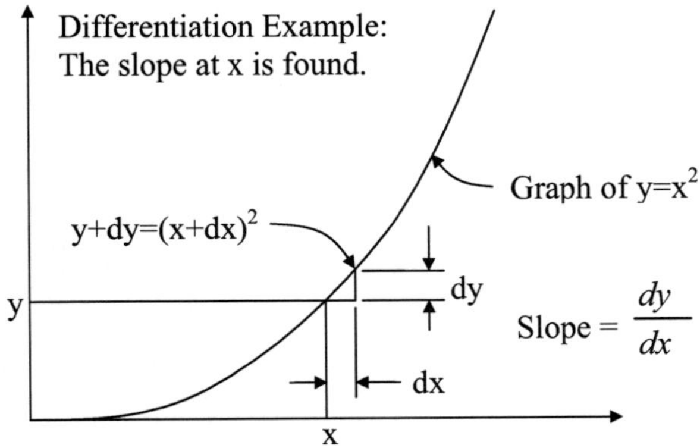

The “d” in the “dy” and “dx”, shown above, is calculus notation that means dy and dx are both very small distances that are as close to zero as possible without actually becoming zero. Solving the above equation for the slope (dy/dx) gives us:

$$y+dy=(x+dx)^2=x^2+2xdx+(dx)^2$$

Substituting x^2 for y we get:

$$x^2+dy=x^2+2xdx+(dx)^2$$

Subtracting x^2 from both sides and dividing both sides by dx gives:

$$dy/dx = 2x+dx$$

As mentioned, the essence of calculus is to examine at the very tiny level and, in doing so, the small error terms go away, giving an exact result. For example, the term “dx” in the above equation is as close to zero as possible. It is much smaller than 2x and can be removed giving us the exact slope of the curve at any x location. We call this the first derivative of the curve $y=x^2$:

$$dy/dx = 2x$$

The slope of the curve $y=2x$ is 2. This is the slope of the slope so we can instead say that the second derivative of $y=x^2$ is 2.

The slope of the curve $y=2$ is zero. This is the slope of the slope of the slope which we call the third derivative of $y=x^2$.

The next chart shows the original function $y=x^2$ along with its first derivative $y=2x$ and its second derivative $y=2$. Notice that we can reverse the derivative process by finding the area under the $y=2$ curve to get the $y=2x$ curve. We can again reverse the derivative process by finding the area under the curve $y=2x$ to get the $y=x^2$ curve. We use the process of integration to go from the top curve to the next curve below. We use the process of differentiation to go from the bottom curve to the next curve above.

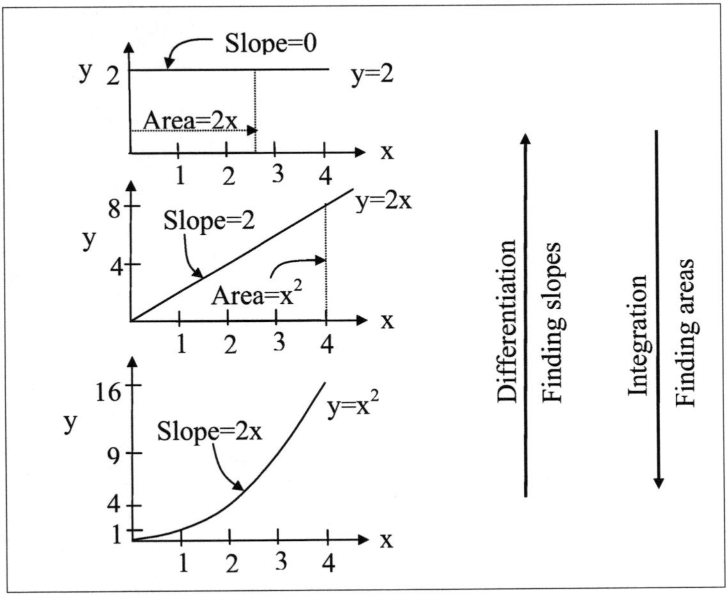

An integration example is shown next where we integrate (find the area of) the function $y=x^2$ between x=1 and x=2.

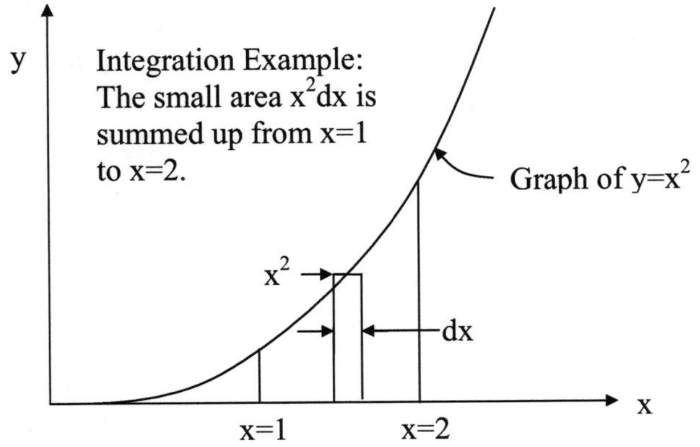

The calculus notation for the sum of the area, A, under the $y=x^2$ curve between the limits of x=1 and x=2 is shown below. The symbol that looks like a vertically stretched "s" is the integration symbol. Think of this symbol as a big "s" that means to sum the small area x^2dx between the range x=1 to x=2. When performing this integration calculation, the area from x=0 to the upper limit x=2 will be calculated, then the area from x=0 to the lower limit x=1 will be subtracted from the upper limit calculation.

$$A=\int_1^2 x^2 dx$$

Calculus books have integration tables for common functions. According to these tables, the function x^2dx integrated to $x^3/3$ giving the result shown below:

$$A=[x^3/3]_1^2=2^3/3-1^3/3=8/3-1/3=2.333$$

Notice in the above equation that the x in the $[x^3/3]$ term is first replaced by the upper limit 2. From this upper limit area, (8/3), the lower limit area (1/3) from x=0 to x=1 is subtracted.

In other Appendix sections, the integration process is used to derive the equation for the area of a circle, for the surface area of a sphere, and for the volume of a sphere.

Pi and Radian Measurement

Radian angle measurement will be used to derive the formula for the area of a circle where 90° is equal to $\pi/2$ radians. An angle of 1 radian occurs when the length of arc of the circle is equal to the circle radius. Here is how it works:

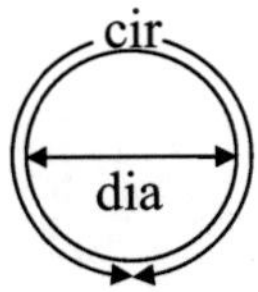

For any size circle in 2-D, non-rotating space:
$\pi = cir/dia = 3.14159\ldots$

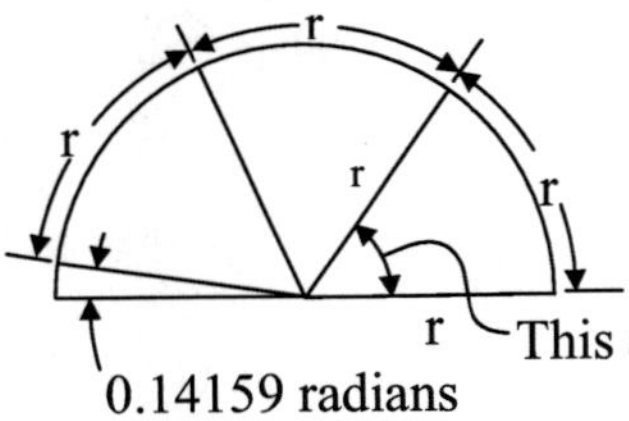

$180° = \pi$ radians
1 radian = $180°/\pi = 57.296$ degrees

The Area of a Circle

Below is shown one quadrant of a circle of radius r. We will use calculus to find the area of this quadrant and then multiply by 4 to get the area of the full circle. The quadrant is further divided into very small slices where the angle of each slice is dθ. In calculus notation, the character "d" in the dθ means that the angle θ is as close to zero as possible without becoming zero. As the angle dθ approaches zero degrees, this small slice approaches the shape of a triangle. This triangle has a height of rdθ and a base of r. The area of this triangle is ½(base∗height) or ½(r∗rdθ) for very small angles. As the angle of dθ approaches zero, the error term in the triangle calculation due to the arc of the circle becomes zero. This is the technique used in calculus to sum all of these very small slices and thus get the exact area of the circle.

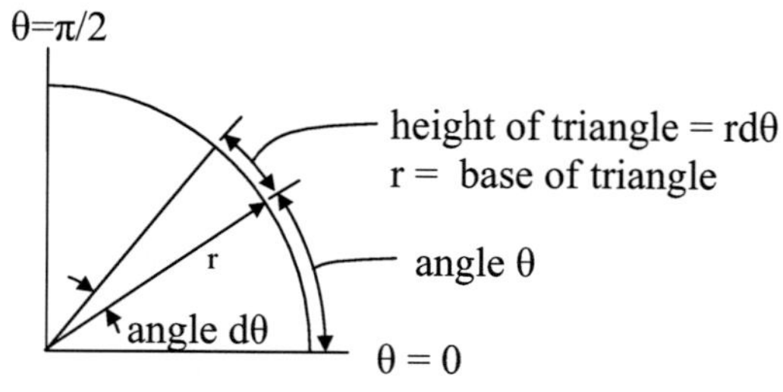

The area of the small triangle slice is:

$$dA = \frac{1}{2}(base * height) = \frac{1}{2} r^2 \, d\theta$$

In calculus notation, the area of the circle is shown in the next equation. The symbol that looks like an "s" stretched in the vertical direction is the integration symbol. This symbol means that the small area dA will be summed from the angle 0 to π/2 to give the full area of the quadrant.

$$A = 4 \int_0^{\pi/2} dA = 4 \int_0^{\pi/2} (1/2)\, r^2 \, d\theta = 2 r^2 \int_0^{\pi/2} d\theta$$

According to the integration rules of calculus, the term dθ sums to π/2 as shown below:

$$A = 2\, r^2 \frac{\pi}{2}$$

The area of a circle is found to be the area of a square of side r, times π:

$$A = \pi\, r^2$$

The Surface Area of a Sphere

Below is shown one quadrant of a sphere that contains one-eighth of the sphere's total surface area. Also shown is the area of a very small region on the surface of the sphere. As the sides of this region approach zero, the area of this region becomes a perfect square. That is because the error terms in the square calculation due to the arc of the surface go to zero. The $rd\theta$ side of the square is the radius r times the smallest non-zero angle, $d\theta$, that is possible. The side $(rd\varphi)\sin(\theta)$ is equal to zero at the top of the sphere and is equal to $rd\varphi$ when the angle θ is at 90 degrees ($\pi/2$ radians). The angle $d\varphi$ is the smallest non-zero angle of φ that is possible. The total area of the sphere is derived below using calculus to sum up the area of each of these small squares over the surface of the sphere.

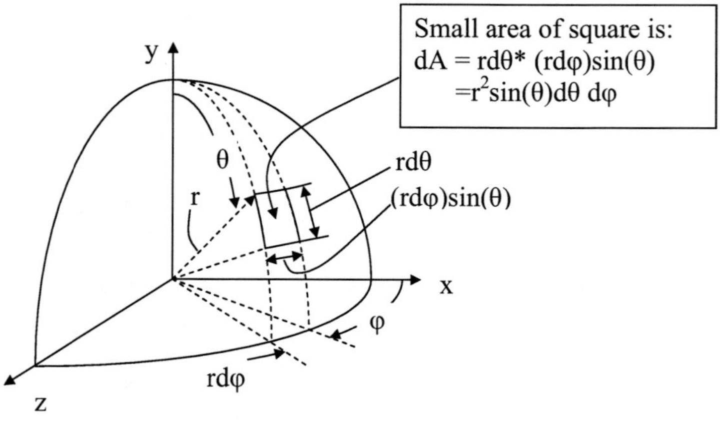

The surface area of the full sphere is written next in calculus notation where $\pi/2$ is the radian angle equivalent of 90°.

$$A = 8 \int_0^{\pi/2} \int_0^{\pi/2} r^2 \sin(\theta)\, d\theta\, d\varphi$$

The symbol that looks like an "s" stretched in the vertical direction is called the integration symbol. The integration symbol means that a function following this symbol is summed over the range expressed by the value directly below this symbol (0) to the value expressed directly above this symbol (π/2). In the equation above, two separate integrations will be done. First sin(θ)dθ will be integrated over the range from 0 to π/2 radians. Next the function dφ is summed over the angle from 0 to π /2 radians.

Calculus books have tables that give the integration result of common functions. According to the integration rules of calculus, integrating the function sin(θ)dθ gives the result -cos(θ). Integrating sin(θ)dθ from 0 radians to π/2 radians gives the equation below. Notice that the cos terms in the brackets have the lower range value subtracted from the upper range value.

$$A = 8 \int_0^{\pi/2} r^2 (-1)[\cos(\pi/2) - \cos(0)] d\theta\, d\varphi$$

Since cos(π/2)=0 and cos(0)=1, the above equation reduces to:

$$A = 8 \int_0^{\pi/2} r^2\, d\varphi$$

Integrating dφ between the limits of 0 to π/2 radians yields:

$$A = 8 r^2 [\pi/2 - 0]$$

The surface area of a sphere becomes 4 times the area of a circle:

$$A = 4\pi r^2$$

Coulomb's Law is a fundamental equation of nature that describes the forces between electrical charges. Note that the surface area of a sphere can be found in Coulomb's Law:

$$F = \frac{q_1 q_2}{\varepsilon_0 4\pi r^2}$$

The Volume of a Sphere

Shown next is a quadrant of a sphere, which is one-eighth of the total sphere volume. Also shown is the volume of a very small region inside of the sphere. As the sides of this region approach zero, the volume of this region becomes a perfect cube since the error terms in the cube calculation go to zero. The "dr" side of the cube is a very small section of the radius that is as close to zero as possible without becoming zero. The $rd\theta$ side of the cube is the radius, r, times the smallest non-zero angle, $d\theta$, that is possible. The side $(rd\varphi)\sin(\theta)$ is equal to zero at the top of the sphere and is equal to $rd\varphi$ when the angle θ is at 90 degrees ($\pi/2$ radians). The angle $d\varphi$ is the smallest non-zero angle of φ that is possible. The total volume of the sphere is now derived using calculus to sum up the volumes of each of these small cubes over the region of the sphere.

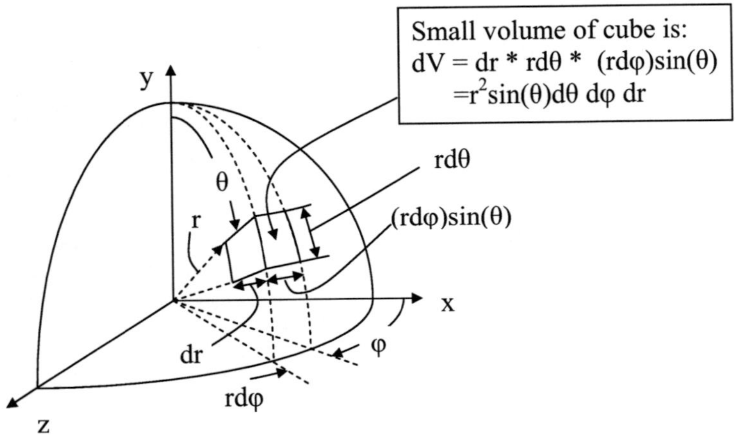

The volume of the full sphere is written below in calculus notation.

$$V = 8\int_0^r \int_0^{\pi/2} \int_0^{\pi/2} r^2 \sin(\theta)\, d\theta\; d\varphi\; dr$$

The symbol that looks like an "s" stretched in the vertical direction is called the integration symbol. The integration symbol means that a function following this symbol is summed over the range expressed by the value directly below this symbol to the value expressed directly above this symbol. In the equation above, three separate integrations will be done. First, sin(θ)dθ will be integrated over the range from 0 to π/2 radians. Next, the function dφ is summed over the angle from 0 to π /2 radians. Finally, the r^2dr is integrated over the radius from 0 to r to give us the full volume of the sphere.

According to the integration rules of calculus, integrating the function $\sin(\theta)d\theta$ gives the result $-\cos(\theta)$. Integrating $\sin(\theta)d\theta$ from 0 radians to $\pi/2$ radians gives the equation below. Notice that the cos terms in the brackets have the lower range value subtracted from the upper range value.

$$V = 8\int_0^r \int_0^{\pi/2} r^2(-1)[\cos(\pi/2) - \cos(0)]d\theta\, d\varphi\, dr$$

Since $\cos(\pi/2)=0$ and $\cos(0)=1$, the above equation reduces to:

$$V = 8\int_0^r \int_0^{\pi/2} r^2\, d\varphi\, dr$$

Integrating $d\varphi$ between the limits of 0 to $\pi/2$ radians yields:

$$V = 8\int_0^r r^2[\pi/2 - 0]\, dr = 4\pi\int_0^r r^2\, dr$$

According to the rules of calculus, integration of the function r^2dr from the limits 0 to radius r gives $r^3/3$. The volume of a sphere becomes:

$$V = \frac{4}{3}\pi r^3 \quad = \quad \frac{r}{3} * (surface\ area\ of\ a\ sphere)$$

A universe of composed of nothing can be thought of as a sphere of zero radius. As the universe comes into existence, the radius first becomes "dr" and then expands to r.

It is interesting that nature has no choice for the volume of a stationary sphere in 3-D space. The constant of nature that determines the relationship between the volume of a sphere and the radius of a sphere cannot be changed. Any 3-D universe must obey this relationship.

The Probability Equation

The laws of probability tell us that an event that has a very small chance of occurring will occur many times given enough time. Remember that earlier we wondered how long it would take for a monkey typing randomly at a keyboard to create Shakespeare's *Hamlet*. If this monkey could type for an infinite amount of time, then the laws of probability require that Shakespeare's *Hamlet* will not only be typed once, but will be created an infinite number of times, along with many other unexpected creations. If nature has this power of creation, then it seems less surprising that the universe, the DNA molecule, and humans exist.

The probability equation can be easily derived. Consider a six-sided die. If we throw this die six times, what is the probability that a single dot will not show up? The probability that we will get a single dot in the first try is 1/6. The probability that we will get something other than a single dot on the first try is (1-1/6) or (5/6). The probability of not having a single dot show up after 6 tosses is:

$$(1-1/6)(1-1/6)(1-1/6)(1-1/6)(1-1/6)(1-1/6) = (1-1/6)^6$$

We can rewrite this as an equation using P_1 as the probability of the event occurring in one try where n is the total number of tries:

Probability event will not happen = $(1-P_1)^n$

Probability event will happen (P_n) = 1- (Probability event will not happen) = $1-(1-P_1)^n$

The probability equation becomes:

$$P_n = 1-(1-P_1)^n$$

Where:

P_n = probability of an event occurring after n tries or n time periods
P_1 = probability of event happening in one try or in one time period
n = number of tries or time periods

Now let's use this equation to find the probability, P_n, of having a single dot come up when the die is tossed 6 times:

$$P_n = 1-(1-1/6)^6 = 0.665$$

For small values of P_1, the probability equation becomes:

$$nP_1 = -2.30\log(1-P_n) \quad \text{Where } P_1<0.005$$

If we set the probability of an event happening at 0.999955, then the above equation becomes very simple:

$$nP_1 = 10 \quad \text{Where } P_1<0.005 \text{ and } P_n = 0.999955$$

The above equation can be used to calculate the number of tries (or time required), n, for low-probability events when we want to have a 99.9955 percent confidence level that the event will take place.

Let's assume that the probability of an event taking place in a one-year period is one in a million. How many years will it take to be 99.9955 percent sure that the event will happen?

The answer is: $n = 10/P_1 = 10/1e\text{-}6 = 10$ million years

The $nP_1=10$ equation shows that as long as the probability of an event taking place is bigger than zero, then this event will eventually take place given enough time. We know that complex life arrived on earth less than 14 billion years after the estimated start of the universe. The early universe was composed of mostly hydrogen atoms and some helium. Humans could not come into existence until the stars had billions of years to convert the hydrogen into the heavier elements such as iron and calcium required for our bodies.

Sometimes nature creates quickly. Given the right conditions, nature can turn small water drops into a complex snowflake within a few minutes.

Olbers' Paradox

The solution to Olbers' Paradox provides us with strong evidence that our universe is not infinite in either size or time, while having a uniform galaxy distribution. If we lived in an infinite universe with approximately uniform galaxy distribution, then our night sky would be glowing as bright as the surface of a star. This would cause the temperature on earth to become thousands of degrees and we could not exist. When we look up at the night sky and see mostly darkness, this is strong evidence that we live in a finite universe.

The next drawing shows a very large section of the universe where we are at the center. Also shown are two rings similar to the rings of an onion. The outer ring has a larger volume than the inner ring and will, on average, contain more galaxies. However, the outer ring is at a greater distance from us, so the luminosity reaching us from the outer ring will be decreased. Mathematical analysis will show us that the luminosity reaching the earth from each ring is the same. When the sum of the luminosity reaching the earth from all possible rings is added up for an infinite radius, the luminosity reaching the earth becomes infinite.

It can be argued that the universe may have dust or other dark matter that will block the light from the outer rings. However, this matter would heat up to thousands of degrees and glow like the rest of the matter in a big cosmic oven. If the universe has not existed for an infinite time, then the universe may still be infinite in size. In this case, we have avoided the cosmic oven since the light from the distant parts of the universe has not had enough time to reach us.

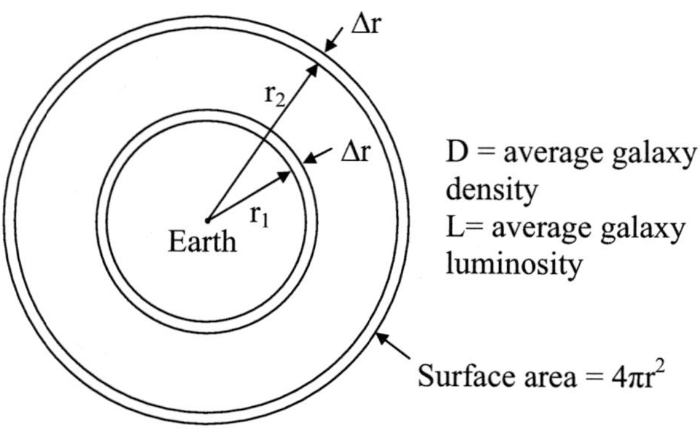

The luminosity of the outer ring as seen from earth is:

$4\pi r_2^2 \Delta r (L*D)/r_2^2$ Where $4\pi r_2^2 \Delta r$ is the volume of the ring

The luminosity of the ring is proportional r^2. Due to the inverse square law of light, the luminosity seen at earth is proportional to $1/r^2$. The r^2 terms cancel each other and the luminosity for any ring becomes:

$$4\pi\Delta r(L*D)$$

If we have 10 rings, then the luminosity seen at earth will be:

$$10*(4\pi\Delta r(L*D))$$

If we have ∞ rings, then the luminosity seen at earth will be:

$$\infty*(4\pi\Delta r(L*D)) = \infty$$

In reality, the inner rings block the light from the outer rings, causing a cosmic oven which will reach the temperature of the average star.

Kinetic Energy (Energy of Motion)

The concept of kinetic energy will later be used to determine the critical mass density of the universe. The critical mass density of the universe is the amount of mass needed in the universe so that the current expansion of the universe will continue forever, without the matter coming back together. Kinetic energy is the energy of an object due to the motion of the object. The concept of kinetic energy can be used to calculate the velocity we must give to an object at the earth's surface in order for this object to escape the gravitational force of the earth and never fall back. In a similar way, we can use the concept of kinetic energy to determine the amount of energy that we must give to any object for the object to completely escape the gravitational attraction of the universe.

Kinetic energy is equal to one-half of the mass of the object multiplied by the square of the velocity of the object:

$$\text{Kinetic Energy} = \frac{1}{2} m\, v^2$$

To see why this is so, consider an object of mass, m, initially at rest. At time zero, a constant force, F, is applied to the object. The object will now experience a constant acceleration (a=F/m) in the direction of the applied force. At time t_1 after the force is applied, the velocity of the object will be v_1:

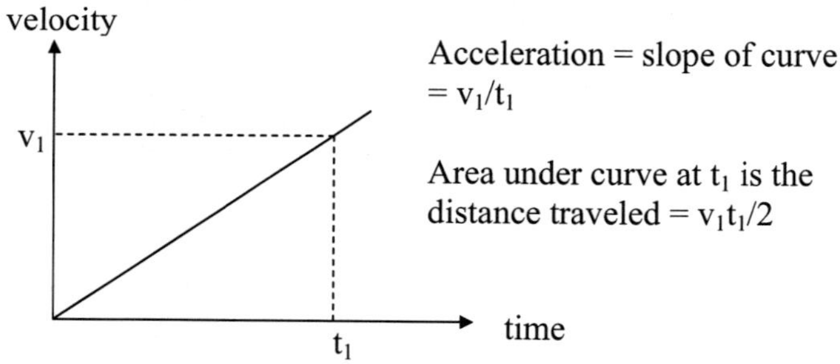

Since energy is a force times a distance and since force is equal to mass times acceleration, we can write:

Kinetic Energy = force*distance = mass*acceleration*distance

Where, in this example:

mass $= m$

$$acceleration = \frac{v_1}{t_1}$$

distance = average velocity*time $= \dfrac{v_1}{2} t_1$

Substituting we get:

$$Kinetic\ Energy = m(\frac{v_1}{t_1}) * (\frac{v_1}{2} t_1) = \frac{1}{2} m v_1^2$$

The Negative Potential Energy of Gravity

The potential energy due to gravity is a negative energy as any standard university physics book will confirm. It will later be shown how the negative energy due to the gravity of the universe is canceled by the positive energy due to the matter in the universe. This results in a universe where the total energy adds to zero. This also explains how the universe can be made from nothing without violating the Law of Conservation of Energy. So let's now investigate the negative energy due to gravity.

Newton's Law of Gravitation describes the force of attraction between two masses, m_1 and m_2, separated by a distance, r:

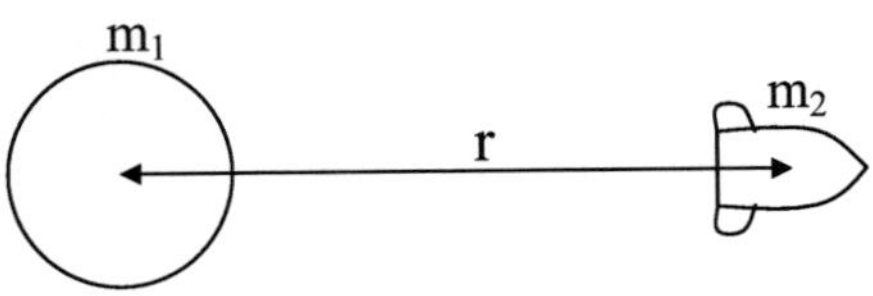

$$F = G \frac{m_1 m_2}{r^2}$$

Where:

F = Force of attraction between mass m_1 and mass m_2
G = Gravitational Constant (6.672×10^{-11} nt-m^2/kg^2)
m_1 = mass of object 1

m_2 = mass of object 2
r = distance between the center of each mass

Energy is a force times a distance:

Energy = force*distance

For example, lifting a large rocket ship 1 foot off the ground requires a large force expending a large amount of energy. Lifting the same rocket 2 feet off the ground requires twice the energy. Lifting a book 1 foot off the ground requires a much smaller force and therefore uses much less energy. As the m_1 and m_2 are separated an infinite distance from each other, the force of gravity becomes zero. Since a force of zero times a distance is always zero, the energy at infinite separation is considered to be zero. As r becomes a finite value, the energy due to gravity becomes negative, as shown below.

Using calculus, we can solve for the energy required to bring object m_2 from infinity (zero energy) to a distance r from object m_2 (negative energy):

$$Energy = \int_{\infty}^{r} F\,dr = \int_{\infty}^{r} G\,m_1 m_2 r^{-2}\,dr = G\,m_1 m_2 \int_{\infty}^{r} r^{-2}\,dr$$

$$Energy = -G\,m_1 m_2 [1/r]\,\Big|_{\infty}^{r} = -G\,m_1 m_2 [(1/r) - 0]$$

$$Energy = -\frac{G\,m_1 m_2}{r}$$

Mathematically we see that the gravitational energy is negative.

This may still seem a bit odd, so let's consider an example of the pendulum shown below.

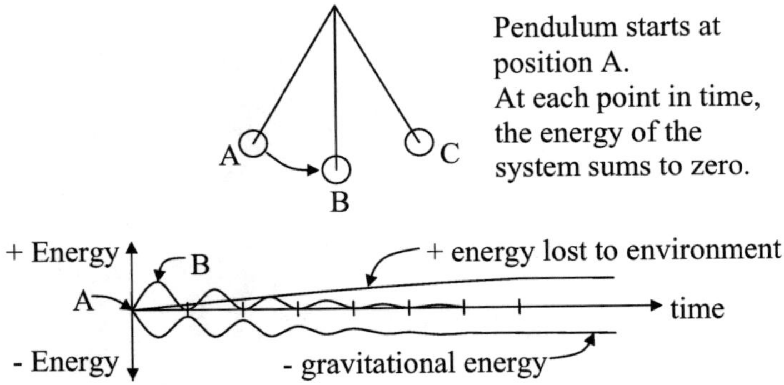

The pendulum is initially released from an elevated position A. We will use position A as the zero energy reference point for the pendulum system. Position A is a point of zero kinetic energy. The gravitational energy of the pendulum mass is less negative when raised and more negative at the bottom of the swing at position B. As the pendulum is released from position A, it increases in positive kinetic energy reaching a maximum at position B.

At position B, the negative energy of gravitation is balanced by the positive kinetic energy. The net energy of the system is still zero at position B as required by the Law of Conservation of Energy. As the pendulum swings to position C, the kinetic energy decreases and the gravitation energy becomes less negative.

As this process continues, the kinetic energy of the system decreases as energy is lost to the environment due to friction. Eventually the pendulum stops at the bottom of the swing. The energy of the system is still zero. When the pendulum stops, the positive energy lost to the environment is exactly balanced by the increased negative energy due to gravity. The graph shows how the negative gravitational energy is exactly balanced by the sum of positive kinetic energy and positive energy lost to the environment.

We will later show how net energy of the universe can be zero due to the balance of positive mass energy by negative gravitational energy.

The Critical Mass Density of the Universe

Hubble's Law shows that the universe is expanding. As we mentioned earlier, the galaxies are separating from each other as the raisins in a loaf of bread separate during baking. The rate of expansion of the universe depends on the mass density of the universe or on how much mass is in an average region of space.

If the mass density is large enough, then the attraction force of gravity between the masses will be large enough to slow down the expansion, eventually causing the universe to collapse inward. If the mass density of the universe is small enough, then the forces of attraction between the masses will be less, allowing the universe to expand to infinity. The critical mass density of the universe is the smallest value that will allow the universe to expand to infinity without collapsing.

Scientists think that we live in a universe that is very close to the critical mass density. The universe is approximately 14 billion years old. If the mass density was more than the critical mass density, then our universe should have collapsed a long time ago. If the mass density was less than the critical mass density, then the universe should have expanded much more than what we observe today. The fact that the universe is still very close to the critical mass density after 14 billion years is strong evidence that it started out being extremely close to the critical value.

We will next calculate the critical mass density for a galaxy of mass m_g to escape the universe of mass M_u. Recall from the previous section on kinetic energy:

$$\text{Kinetic energy of galaxy} = ½\, m_g v^2$$

Also recall from the previous section that the negative potential gravitational energy of the galaxy is:

$$\text{Gravitational energy of galaxy} = -GM_u m_g/r$$

For the galaxy to escape the gravitational force of the universe, the galaxy must have positive kinetic energy that is equal to the negative gravitational potential energy:

$$(½)m_g v^2 = GM_u m_g/r$$

Notice that m_g cancels from both sides of the equation:

$$(½)v^2 = GM_u/r$$

From Hubble's Law, the velocity of the galaxy is:

$$v=Hr \quad \text{Where } H = \text{Hubble's constant}$$

Substituting we get:

$$½(Hr)^2 = GM_u/r$$

The mass of the universe, M_u, is equal to the volume of a sphere of radius r times the critical mass per unit volume:

$$M_u = (4/3)\pi r^3(\text{Crit\_den})$$

Where Crit_den = critical mass density of the universe

Substituting for M_u we get:

$$\tfrac{1}{2}(Hr)^2 = G((4/3)\pi r^3(\text{Crit\_den}))/r$$

Canceling r from both sides we get:

$$\tfrac{1}{2}H^2 = G(4/3)\pi(\text{Crit\_den})$$

Solving for the critical mass density of the universe we get:

$$\text{Crit\_den} = 3H^2/(8\pi G)$$

This equation will be used in the next section to show that the positive mass energy in the universe can be exactly balanced by the negative gravitational potential energy in the universe. This results in a universe in which the total energy equals zero.

How the Total Energy of the Universe Sums to Zero

In this section, we will show that the total positive mass energy of the universe is balanced by the total negative gravitational potential energy of the universe, resulting in zero total energy. We will next investigate this mathematically. When doing this, we will arrive at Einstein's famous equation, $E=mc^2$, purely by considerations of the behavior of the universe on a large scale.

In a previous section, it was shown that the gravitational potential energy of a mass, m, at a distance, r, from mass M_u can be written:

$$E=-GM_u m/r \quad \text{Equation A}$$

In Equation A, let's consider M_u to be the mass of the universe and consider the gravitational energy of a mass, m, that sits at an average "effective" distance from M_u as half the Hubble radius, R, of the universe. The Hubble radius of the universe is:

$$R=c/H$$

Where:

c = speed of light
H = Hubble's Constant

Setting r at half the Hubble radius gives:

$$r=c/2H \quad \text{Equation B}$$

Substituting Equation B into Equation A gives:

$$E=-GM_u m(2H/c) \quad \text{Equation C}$$

M_u is the critical mass density of the universe multiplied by the volume of a sphere of Hubble radius R=c/H:

$$M_u=(4/3)\pi(c/H)^3*Crit\_den \quad \text{Equation D}$$

In a previous section we showed that the critical mass density of the universe is:

$Crit\_den = 3H^2/(8\pi G)$ Equation E (See previous section for proof.)

Substituting Equation E into Equation D gives:

$$M_u=(4/3)\pi(c/H)^3*3H^2/(8\pi G) = c^3/2HG \quad \text{Equation F}$$

Substituting Equation F into Equation C gives:

$$E=-G(c^3/2HG)m(2H/c) \quad \text{Equation G}$$

Canceling the terms in Equation G gives:

$$E=-mc^2$$

This amazing result shows that the negative gravitational potential energy is equal to the energy of the mass according to Einstein's famous equation.

Let's consider what this result really means. It says that the negative energy due to the gravity of mass m is balanced by the positive mass energy. These two energies cancel each other out, giving a net energy of zero for the universe.

To see how this works, let's imagine that the mass, m, is a rocket ship. Let's say we have a special engine that can convert mass energy into gravitational potential energy. We will use the rocket's mass as fuel to escape the universe. If the rocket ship can convert half of its mass into gravitational potential energy, then the rocket will be one-half the mass but will have doubled its distance from the universe. The rocket is now twice as far from the mass of the universe, and the strength of gravity has been reduced by one-fourth. At this reduced strength of gravity, the rocket can again burn one-half of its remaining mass and again double its distance from the universe. Each time the rocket burns one-half of its remaining mass, it will double its distance from the universe. Since the force of attraction is reducing, the amount of mass needed to double the distance is reduced for each burn. This process continues until the rocket has zero remaining mass at an infinite separation. At this infinite separation, the gravitational energy has also become zero.

The rocket ship fuel burns are shown graphically below:

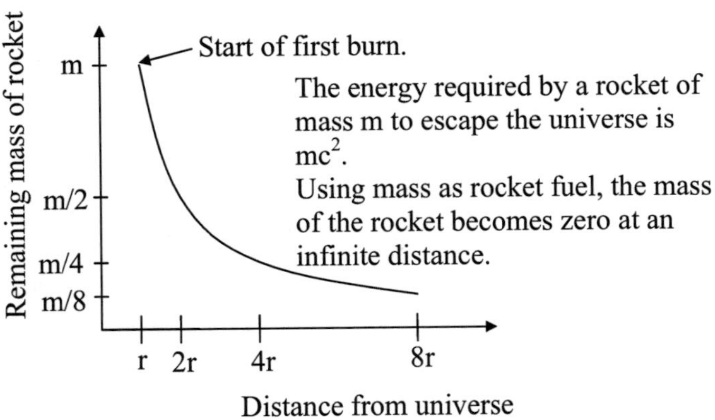

This shows how the positive mass energy can be exactly balanced by the negative gravitational potential energy of the universe.

Calculation of the Gravitational Energy in the Universe

In this section, we will use integral calculus to determine the gravitational energy of the universe. We will assume a spherical universe with the critical mass density that was calculated in a previous section. This critical mass density is repeated below:

$$\text{Critical\_density} = CD = 3H^2/(8\pi G)$$

Where:

H= Hubble's constant
G= Gravitational Constant

We will next calculate the gravitational energy in a thin outer ring of the universe, which is similar to the outer ring in an onion. We will then use integration to sum the energy in all of rings from a radius of 0 to a radius of r.

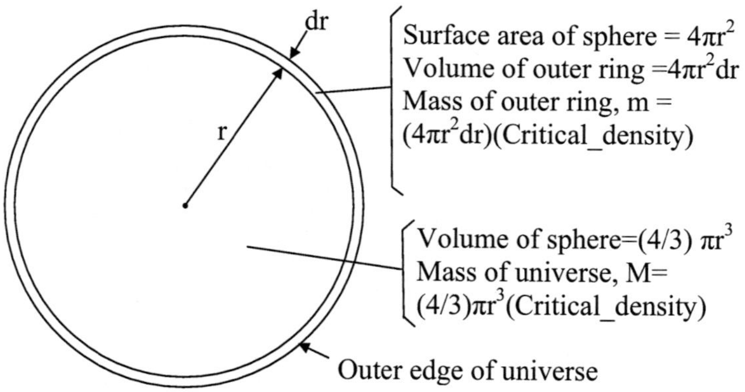

The gravitational energy in the outer ring of the universe is:

$$E=-G\frac{Mm}{r}=-G\frac{[(4/3)\pi r^3(CD)]*[4\pi r^2dr(CD)]}{r}$$

Where:

M = mass of universe
m = mass of outer ring
CD = critical mass density of universe

Simplification of the above equations gives us the energy of the outer ring:

$$E=-\frac{16}{3}\pi^2G(CD)^2r^4dr$$

Using integration to sum the energy, E_g, contained in all of the rings gives us:

$$E_g=-\frac{16}{3}\pi^2G(CD)^2\int_0^r r^4dr=-\frac{16}{3}\pi^2G(CD)^2\frac{r^5}{5}\Big|_0^r$$

Subtracting the upper limit of integration from the lower limit gives us:

$$E_g=-\frac{16}{15}\pi^2G(CD)^2r^5$$

Substituting the value of the critical density gives us:

$$E_g = -\frac{16}{15}\pi^2 G\left(\frac{3H^2}{8\pi G}\right)^2 r^5$$

Simplifying the above equation gives us the gravitational energy of the universe:

$$E_g = -\frac{3H^4 r^5}{20G}$$

Where:

H = Hubble's Constant
G = Gravitational Constant
r = radius of universe

In the next section, we will use this result to determine the radius of the universe based on the energy balance of the positive mass energy with the negative gravitational energy.

The Radius of the Universe Based on Energy Balance

In the previous section we derived the gravitational energy of the universe as:

$$E_g = -\frac{3H^4 r^5}{20G} \quad \text{Equation A}$$

Where:

H = Hubble's Constant
G = Gravitational Constant
r = radius of universe

By setting the mass energy of the universe equal to the above gravitational energy of the universe, we can determine the radius of the universe based on the balance of positive mass energy by negative gravitational energy.

The positive mass energy, E_u, of the universe can be calculated as follows:

$$E_u = M_u c^2 \quad \text{Equation B}$$

Where:

M_u = mass of universe
c = speed of light

M_u is the volume of the universe multiplied by the critical density of the universe:

$$M_u = \left(\frac{4}{3}\pi r^3\right)\left(\frac{3H^2}{8\pi G}\right) = \frac{H^2 r^3}{2G} \quad \text{Equation C}$$

Substituting Equation C into Equation B gives us:

$$E_u = \frac{H^2 r^3}{2G} c^2 \quad \text{Equation D}$$

By setting the mass energy of Equation D equal to the magnitude of the gravitational energy in Equation A, we get:

$$\frac{H^2 r^3}{2G} c^2 = \frac{3 H^4 r^5}{20G}$$

Solving the above equation for the radius of the universe gives us:

$$r = \sqrt{\frac{10}{3}} \left(\frac{c}{H} \right)$$

Where:

c = speed of light
H = Hubble's Constant

We will use this result in the next section to calculate the escape velocity of the universe.

Why Our Universe May Be a Black Hole

The escape velocity of the universe is the speed that an object at the edge of the universe must have in order to escape the gravitational influence of the mass in the universe. If the outward speed of the object is equal to or greater than the escape velocity, then the object will escape from the universe. If the speed is less than the escape velocity, then the object will fall back into the universe. If the escape velocity is greater than the speed of light, then it is possible that we live in a universe that is a black hole—not even light can escape.

We will now calculate the escape velocity for an object at the edge of a spherical universe.

By setting the kinetic energy of the object at the outer edge of the universe equal to the gravitational energy of the object, we can derive the equation for the escape velocity:

$$\frac{1}{2} m\, v^2 = \frac{G\, M_u\, m}{r}$$

Where:

m = mass of moving object at the edge of the universe
M_u = mass of universe
r = radius of universe
G = Gravitational Constant

Solving for the velocity, we get:

$$v = \sqrt{\frac{2GM_u}{r}} \quad \text{Equation A}$$

In a previous section we determine the mass of the universe to be:

$$M_u = \frac{H^2 r^3}{2G} \quad \text{(See Equation C in previous section.)}$$

Substituting the above equation into Equation A, we get:

$$v = \sqrt{\frac{2G(\frac{H^2 r^3}{2G})}{r}} = Hr \quad \text{Equation B}$$

In the previous section, we derived the radius of the universe based on energy balance as:

$$r = \sqrt{\frac{10}{3}}\left(\frac{c}{H}\right)$$

Substituting the above equation into Equation B gives us the escape velocity:

$$v = \sqrt{\frac{10}{3}}\, c = 1.83c$$

Since the escape velocity of an object at the edge of the universe exceeds the speed of light, we may live in a black hole.

The Alice and Bob Paradox

When we view the distant night sky through a high-powered telescope, we are seeing into the past. We see the light as it was billions of years ago. With a very powerful telescope, we should be able to see back in time close to the beginning of the universe.

Alice and Bob are two observers looking back in time. Alice is at the North Pole and Bob is at the South Pole. Both have powerful telescopes. They both look straight up and back in time toward the birth of the universe. Yet, they are looking in opposite directions! How can this be? The origins that they see are at the opposite ends of the sky. How can the beginning of the universe be spread out in all directions? Shouldn't it be a single point in space? Could the universe have formed by an implosion rather than an explosion? See the following picture.

The Alice and Bob Paradox

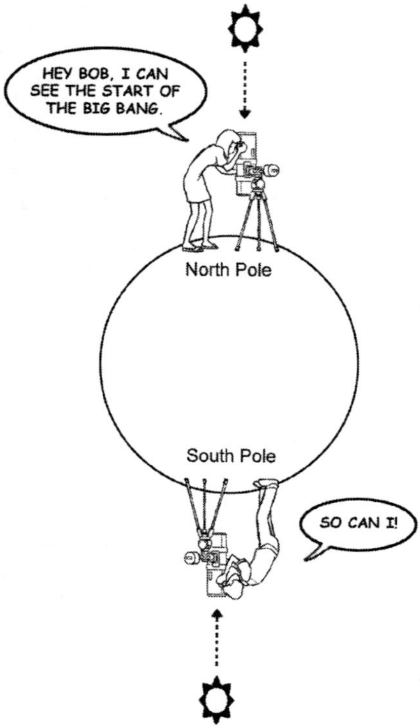

Both Alice and Bob are looking back in time with their powerful telescopes and see the early light from the Big Bang. But they are each looking in opposite directions. If the Big Bang originated at a point in space, then how can this be?

Bibliography

Gravity as negative energy:

Arfken, G., Griffing, D., Kelly, D. and J. Priest. *University Physics*. San Francisco: Academic Press, Inc., 1984. p. 158.

Hawking, Stephen. *A Brief History of Time.* New York: Bantam Books, 1996. p. 133.

Resnick, Robert and David Holliday. *Physics For Students of Science and Engineering.* New York: John Wiley &Sons, Inc., 1960. pp. 341-342.

The energy balance between mass and gravity:

Hawking, Stephen. *A Brief History of Time.* New York: Bantam Books, 1996. p. 133.

Potter, Franklin, and C. Jargodzki. *Mad About Modern Physics.* New Jersey: John Wiley & Sons, Inc., 2005. p. 275.

Wolff, Milo. *Exploring the Physics of the Unknown Universe.* Manhattan Beach: Technotran Press, 1994.

The Heisenberg Uncertainly Principle and/or virtual particles:

Cutnell, John D. and Kenneth W. Johnson. *Physics*, Fourth Edition. New York: John Wiley & Sons, Inc., 1998. pp. 901-905

Hey, Tony, and Patrick Walters. *The Quantum Universe.* New York: Cambridge University Press, 1990.

Evolution of the chemical elements:

Emsley, John. *Nature's Building Blocks*. Oxford: Oxford University Press, 2003.

Sagan, Carl, and Iosif S. Shklovskii. *Intelligent Life in the Universe.* San Francisco: Holden-Day, Inc., 1966.

The start of life:

Miller, Stanley. "A Production of Amino Acids Under Possible Primitive Earth Conditions." *Science* Vol. 117 (1953): pp. 528-529.

A very readable biology text is:
Clark, David and Lonnie Russell. *Molecular Biology* made simple and fun. St. Louis: Cache River Press, 2000.

Calculus:

Smail, Lloyd L. *Analytic Geometry and Calculus*. New York: Appleton Century Crofts, Inc., 1953.

^**